Project AIR FORCE

T0150307

MOTIVATED METAMODELS

SYNTHESIS OF CAUSE-EFFECT REASONING AND STATISTICAL METAMODELING

Paul K. Davis

James H. Bigelow

Prepared for the
United States Air Force

RAND

The research reported here was sponsored by the United States Air Force under Contract F49642-01-C-0003. Further information may be obtained from the Strategic Planning Division, Directorate of Plans, Hq USAF.

Library of Congress Cataloging-in-Publication Data

Davis, Paul K., 1943–
    Motivated metamodels : synthesis of cause-effect reasoning and statistical
metamodeling / Paul K. Davis, James H. Bigelow.
      p. cm.
    "MR-1570."
    Includes bibliographical references.
    ISBN 0-8330-3319-0 (pbk.)
    1. Computer simulation. 2. Decision making. I. Bigelow, J. H. II. Rand Corporation.
III.Title.

QA76.9C65D395 2003
003.3—dc21

                          2002155700

Published 2003 by RAND
1700 Main Street, P.O. Box 2138, Santa Monica, CA 90407-2138
1200 South Hayes Street, Arlington, VA 22202-5050
201 North Craig Street, Suite 202, Pittsburgh, PA 15213-1516
RAND URL: http://www.rand.org/
To order RAND documents or to obtain additional information, contact Distribution
Services: Telephone: (310) 451-7002; Fax: (310) 451-6915; Email: order@rand.org

Preface

This monograph was prepared as part of a project on multiresolution modeling for the United States Air Force Research Laboratory (AFRL). The project's goal is to extend the theory and application of techniques for multiresolution, multiperspective modeling (MRMPM). MRMPM is of interest in many disciplines because models exist at different levels of detail and are written from different viewpoints. Sometimes it is necessary to understand their relationships to each other. This study is correspondingly broad in its implications. It is specifically relevant to an ongoing AFRL program that is developing decision aids for effects-based operations (EBO). Decision aids should, where possible, depend only on relatively simple, understandable, fast-running, and agile models that are also easy to maintain. Metamodeling methods can create such models based on experiments with more detailed models, such as TAC THUNDER, STORM, and BRAWLER. However, such models can have subtle but serious shortcomings. In this study we explore some of the issues in depth for a particular example. We then suggest ways to improve the quality of metamodeling by striking a synthesis between statistical and more phenomenological approaches. The work is intended primarily for analysts and those concerned with analytical methodology being used by the Air Force.

Our research was conducted in RAND's Project AIR FORCE. Project AIR FORCE, a division of RAND, is the United States Air Force's Federally Funded Research and Development Center (FFRDC) for studies and analyses. It provides the Air Force with independent analyses of policy alternatives affecting the development, employment, combat readiness, and support of current and future aerospace forces. Research is primarily performed in four programs: Aerospace Force Development; Manpower, Personnel, and Training; Resource Management; and Strategy and Doctrine. This research, however, was performed on a divisionwide level. Comments are welcome and should be addressed to the senior author (e-mail: pdavis@rand.org).

Contents

vi

Figures

Tables

Summary

Background

Simple, low-resolution models are needed for high-level reasoning and communication, decision support, exploratory analysis, and rapidly adaptive calculations. Analytical organizations often have large and complex object models, which are regarded as reasonably valid. However, they do not have simpler models and cannot readily develop them by rigorously studying and simplifying the object model. Perhaps the object model is hopelessly opaque, the organization no longer has the expertise to delve into the model's innards, or there simply is not enough time to do so. One recourse in such instances is statistical metamodeling, which is often referred to as developing a response surface. The idea is to emulate approximately the behavior of the object model with a statistical representation based on a sampling of base-model "data" for a variety of test cases. No deep knowledge of the problem area or the object model is required.

Unfortunately, such statistical metamodels can have insidious shortcomings, even if they are reasonably accurate "on average." This monograph describes some of those shortcomings and proposes a way (*motivated metamodeling*) to do better, which amounts to drawing upon an approximate understanding of the phenomena at work (i.e., upon approximate theory) to suggest variables for and perhaps the analytical form of the metamodel. This approach is hardly radical, but it is quite different from what happens in normal statistical metamodeling. The quality of metamodels can sometimes be greatly improved with relatively modest infusions of theory.

Shortcomings of Statistical Metamodels

We have focused on four problems with pure statistical metamodels, problems that may be either minor or major, depending on context and the statistical methods used. The problems are

- *Failure to tell a story.* High-level decisionmakers avoid making decisions based on models that they do not fully understand, especially when they know that uncertainties abound. They often seek a robust *logic* for their choices, a logic that makes sense to them and can be explained to others.

They may also value an analytic understanding, such as a "roughly right" formula that displays issues transparently.

- *Failure to represent the multiple-critical-component problem.* If the real-world system being analyzed with a model will fail if *any* of several components fail (which implies a type of nonlinearity), then statistical metamodels will commonly fail to predict that, instead predicting that a weakness in one component can be compensated by greater strength of another.

- *Implications for resource allocation.* The relative "importances" ascribed to input variables in the process of statistical metamodeling with garden-variety methods can be an artifact of the statistics package and the order in which calculations are conducted. They may be a very poor basis for decisions about allocating resources.

- *Shortcomings in the presence of an adversary.* A statistical metamodel may emulate base-model results reasonably well on average, but fail badly in what appear to be obscure corners of the input space (see Figure S.1 for a simple example). These corners, however, may be unusually important if an adversary is seeking to exploit the system's weaknesses.

Exploration of How to Do Better

With these issues in mind, we have explored ways to improve metamodeling, relative to a baseline of simple statistical models, by using an understanding of the problem to inform the metamodel's form—and to do so without invoking advanced statistical methods that are currently difficult for analysts, other than professional statisticians, to understand and apply. After developing a series of hypotheses based on general reasoning and our past experience, we conducted a series of experiments to help sharpen our understanding of issues. We began with a well-documented version of a relatively small but complex model for a particular military problem (using aircraft and other long-range fires to halt an invading army). We used that as the object model for experimentation, that is, as a surrogate for a more typical large and complex object model. We then generated model outputs for a broad range of cases and a series of metamodels. The first were "naïve statistical metamodels" that relied only on commonly used tools and methods. We then developed additional metamodels informed by increasing amounts of "theory." The experiment was by no means rigorous, since we were familiar with the problem and object model, but we did our best to distinguish among readily available physical insights, insights that required more thoughtful analytical work and insights that might not be available to a new person on the scene without weeks of work or excellent model documentation.

RAND*MR1570-S.1*

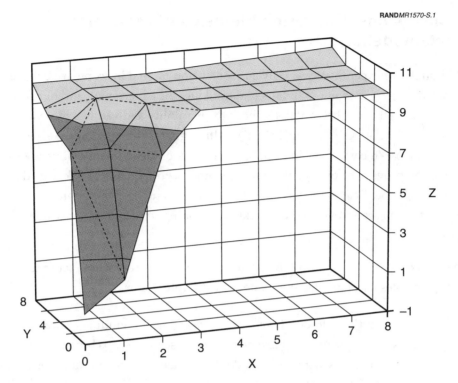

NOTE: A linear metamodel would be a plane surface with values of about 10–11, tipped slightly to reflect data near the origin.

Figure S.1—Metamodeling When the Object Model Has Unusual Behavior in a Corner

The results were interesting and frequently not what we had anticipated. First, for this problem uninformed statistical metamodeling sometimes did better than we had expected in terms of mirroring the object model's *average* performance— especially when we used various techniques well known to statisticians who do metamodeling. Second, some of the simple injections of theory that we hoped would work well did not reduce average error significantly, which was disappointing until we internalized the point that minimizing average error was not really the key issue; rather, the issue was to mitigate the other shortcomings noted above. The motivated metamodels did indeed greatly improve "the story," avoid errors associated with the nonlinearities introduced by critical components, improve the quality of importance estimates for use in resource allocation, and deal better with exploitable vulnerabilities. Sometimes, they also improved average accuracy of the metamodel, but that was less important.

Conclusions: Suggested Elements of Motivated Metamodeling

Although our experiment was only a first step and limited in scope, we believe—based on a combination of fundamental thinking, the experiment, and our past experience—that our basic notions about the feasibility and value of motivated metamodeling are valid (Chapter Four). Although it is difficult to prove, we believe that the "theory" needed to suggest the form of the motivated metamodel will often prove to be obtainable with only modest-to-moderate work (e.g., days or weeks, rather than many months). Good system engineers and policy analysts, after all, are accustomed to quickly developing reductionist constructs of problems.

We also concluded that the following admonitions are a reasonable set of tentative principles for motivated metamodeling (Chapter Four):

1. *Identify the critical components for the problem of interest.* Finding critical components will require some subject-area expertise or perhaps broad engineering talents, but may not require in-depth mathematics or careful deconstruction of the object model.

2. *Postulate a structural form* for the metamodel based on simplified physical reasoning, concern for the critical-component issue, and dimensional analysis. This form may, for example, have a product of factors representing the critical components (i.e., it may be decidedly nonlinear).

3. *Identify important branches.* If the object model has nonlinearities such as Solution = MIN[Solution 1, Solution 2], consider reflecting the same nonlinearity in the metamodel, with different metamodels for the two solution cases.

4. *Identify natural composite variables (aggregation fragments) to use as variables of regression analysis* (e.g., one might expect that variables x, y, and z would ordinarily enter the problem only in the combination xy/z).

5. *Build in fudge factors.* Compensate for imperfections of the postulated form by building in unknown coefficients and error terms. If these turn out to be nonzero but small, they will improve accuracy with little sacrifice of story.

6. *Where applicable, "game" the problem to identify domains of special interest.* These would include worst-case adversary strategies and variables that are strategically correlated. The result may affect the experimental design for sampling base-model data or the form of the model itself.

Conclusions About Model Validation and Documentation

A spinoff of our research was to suggest ideas for model validation and documentation that we have not seen emphasized previously. We recommend that (see Chapter Four):

- Future guidance on model validation should highlight the need to test whether the model correctly represents critical-component and adversary-process issues.

- Guidelines for documenting models should admonish authors to identify critical components and adversary-process issues and to suggest approximate analytical forms, either as an aid to understanding or as the basis for possible motivated metamodeling.

Next Steps

The problem that we examined in this research was narrow. Significantly more thinking will be necessary to extend the ideas to other classes of problems—for example, when the object model is itself imperfect in some respects and other sources of knowledge exist about truth; when only a small number of data points are available to represent behavior of the object model; when the object model is stochastic; when it is important to fine-tune the motivated metamodel for the purposes of a particular analysis; or when analysts need computerized aids (including some of those used in data mining) to help them discover physical insights that could then be used to motivate metamodeling. Such topics would be appropriate subjects of further research. So also is it important in future work to assess how best to use more advanced statistical methods together with phenemenological information.

Acknowledgments

We would like to acknowledge our appreciation to colleagues Brian Williams and Carl Rhodes for thoughtful and informative reviews that clarified the work and added additional pointers to relevant statistical literature.

Acronyms and Abbreviations

EBO	Effects-based operations
IHVR	Integrated hierarchical variable-resolution modeling
MRM	Multiresolution modeling
MRMPM	Multiresolution, multiperspective modeling
SPIE	International Society for Optical Engineering

1. Introduction

Objective

Our purpose in this monograph[1] is to suggest principles for what we call
motivated metamodeling. A metamodel is a relatively small, simple model that
approximates the behavior of a large, complex model. A common way to
develop a metamodel is to generate "data" from a number of large-model runs
and to then use off-the-shelf statistical methods without attempting to
understand the model's internal workings. We describe research illuminating
why it is important and fruitful, in some problems, to improve the quality of
such metamodels by using various types of phenomenological knowledge. Thus,
we strive for a synthesis of techniques across the disciplines of statistics and
computer science on the one hand and science and engineering on the other. The
basic ideas are simple, but current practice is very different from what we
suggest—in part because of disciplinary parochialism. We therefore discuss the
simple ideas in some detail and illustrate them with an experiment that proved
quite useful in sharpening our thinking.

The outline of the study is as follows. In the remainder of this chapter, we lay
out the context for our work: continuing research in the theory of
multiresolution, multiperspective modeling (MRMPM); increased interest of the
analytical community in statistical metamodeling; common practices for
conducting such metamodeling; our initial concerns about these standard
practices; and our notions about how to do better. Chapter Two describes our
research approach, which included an in-depth experiment to sharpen our
understanding of issues. Chapter Three describes the results of the experiment
and lessons learned. Chapter Four draws from the experiment and more general
reasoning to provide conclusions and recommendations; it also discusses
obstacles to acceptance to our suggestions and next steps for research. The
appendices document our work in some detail, including a description of the
specific problem and models used in our experiment.

---

[1]Interim results were presented in conferences. See Davis and Bigelow (2001) and Bigelow and
Davis (2002).

Study Context

Background on Multiresolution, Multiperspective Modeling

We first review ideas discussed in earlier work.[2] The section can be skipped by those already familiar with MRMPM and the need for low-resolution models.

Single Models. Multiresolution, multiperspective modeling is a cutting-edge issue in modeling and simulation.[3] MRMPM has a fundamental role in a wide range of subjects, including the development of military or political-military decision aids (such as those for effects-based operations—EBO),[4] defense planning,[5] and the development of machine intelligence.[6]

A multiresolution model is designed to be used at two or more alternative levels of resolution. That is, a user may enter detailed inputs or a smaller number of lower-resolution inputs. In Figure 1.1, for example, the model has two outputs, A and B. The user may enter the lowest-level inputs, of which there are 17 (the X's). Alternatively, he may choose to enter the problem at a higher level (for example, inputting only 4, 10, or 14 variables), depending on his needs. This flexibility is achieved at the level of code by putting in "switches" (see Appendix A).

Preferably, multiresolution models or families of models are *integrated*, by which we mean that the inputs to a lower-resolution model are related in a straightforward and well-understood way to outputs of a higher-resolution model. The ideal here is integrated hierarchical variable-resolution modeling (IHVR modeling), where we use "variable resolution" as synonymous with multiresolution.[7] Figure 1.1 has this character: the two trees are distinct, as are

---

[2]See Davis (1993), Davis and Hillestad (1993), or Davis and Bigelow (1998).

[3]National Research Council (1997). The USAF Scientific Advisory Board reinforced the matter in recent suggestions to the Air Force Research Laboratory (AFRL), noting the difficulties of integrating both vertically and horizontally among components in model systems. The problem is even more acute when dealing with systems of systems. AFRL has been supporting related research in recent years, much of which has been presented in yearly conferences and reported in *Proceedings of the SPIE* (e.g., Sisti and Farr, 1999). Most recently, multiresolution modeling (MRM) and the closely related issue of model abstraction were highlighted throughout a recent international Dagstuhl Seminar, "Grand Challenges for Modeling and Simulation," August 26–30, 2002, Dagstuhl, Germany.

[4]Davis (2001b)

[5]Davis (2002b)

[6]Meystel (1995), NIST (2001), Davis (2001c), and Meystel and Albus (2002).

[7]A related topic is multimodeling, which is discussed in Chapter 8 of Fishwick (1995). A multimodel is a model composed of other models in a network or graph. The various models may or may not vary in resolution. Zeigler, Praenhofer, and Kim (2000, Chapter 13) discuss families of models and model-abstraction issues

RAND*MR1570-1.1*

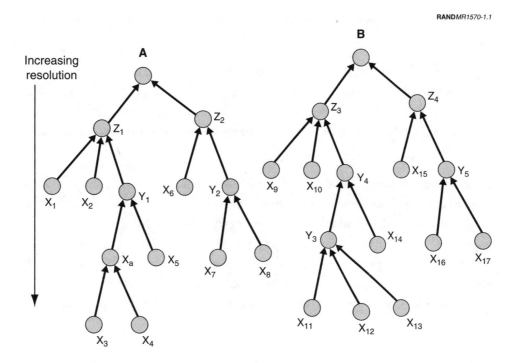

Figure 1.1—Integrated Hierarchical Variable-Resolution (Multiresolution) Modeling

the branches within each of them. This means that a given intermediate variable depends only on the variables below it in its branch. For example, to calculate Z_1, we need only know X_1, X_2, X_3, X_4, and X_5, but not X_6, ... X_{17}. That is, Z_1 can be calculated independent of Z_2, Z_3, and Z_4 and their determinants.

Another way to look at this is that the portion of the tree feeding upward into Z_1 can be regarded as an optional and modular subroutine. This module is not needed elsewhere in the model; it could be present but used only when a switch is set appropriately, or it could be set aside on the shelf as a separate program to be "connected up" only in model runs when it is needed.

Families of Models. Introducing too much such flexibility into a single model results in a good deal of complication. An alternative is to use a *family* of models to accomplish the same purpose: a given member of the family may be designed for one or more levels of resolution, while another member will be designed for other levels. It is often useful for a given model to have the flexibility to go up or down a notch or two in detail, but we would ordinarily not want a single model to go too far in that respect. For example, it would be pointless to include engineering-level details of a given aircraft's radar processor in a theater-level model. It would suffice to attribute engagement ranges to the aircraft—or perhaps engagement-range functions dependent on the types of target (e.g., a normal enemy fighter versus a stealthy one). Something even simpler might be appropriate.

Again, however, the ideal is to have a hierarchical family such that one could—from time to time—use the detailed model to inform or even calibrate parameter values of a higher-level model. The hierarchical ideal is convenient because such calibrations can be made narrowly, without taking into account everything else going on.

Multiperspective Models. Models within a family (or modes within a single model) may also differ in the perspective with which they describe a problem—i.e., the choice of independent variables. This is somewhat akin to differences of representation in physics.[8] In military problems such flexibility is quite important because perspective varies depending, for example, on whether one is in combat arms or logistics, focusing on actual combat or on C4ISR, and so on.[9] Multiresolution, multiperspective modeling, then, includes both aspects of flexibility.

Why Models at Different Levels of Detail Are Needed

Models of different resolution have different strengths and weaknesses. Detailed models are particularly valuable for representing explicitly the underlying phenomena. Good models of this type are an embodiment in mathematics or computer programs of our deepest knowledge of the subject in question. Deep knowledge, however, is not the only important knowledge (or even the *most* important knowledge). Further, much of our knowledge of the world comes from low-resolution sources. In military affairs, this may take the form of historical accounts from the summary perspective of wings or entire commands, rather than, say, the aircraft in a given squadron. In the civilian world, it may take the form of, say, the safety record of an airline as reported in database that contains no information other than the frequency of accidents.

In this study, our focus is on low-resolution models, which are important for a variety of reasons, as discussed below.

Cognitive Needs. Insightful strategy-level analysis and decision support typically require relatively simple models. The most fundamental reason is cognitive: decisionmakers need to *reason* about their issues and inject their own judgments

---

[8]Haimes (1998) refers to this consideration with the terminology "holographic models." Fishwick (1995) also mentions the importance of perspective in discussing what he calls multimodels.

[9]C4ISR is command, control communications, computing, intelligence, surveillance, and reconnaissance.

and perspectives. They need to construct coherent stories that are convincing to themselves and to others. This implies abstracting what may be a very complex problem to a relatively small number of variables or cognitive chunks (e.g., 3–5 rather than 10s or 100s) and somehow focusing on the appropriate variables and cause-effect relationships. "Appropriateness" depends on the context of decision, because simplifications gloss over issues that may be important in other contexts.

Exploratory Analysis Under Uncertainty. Another fundamental reason for using low-resolution models is that strategy-level problems may be characterized by massive uncertainty in many dimensions.[10] The appropriate way to address such problems is often *exploratory analysis,*[11] in which one examines issues across the entire domain of plausible initial states.[12] That is quite different from sensitivity analysis around some reasonably good baseline state. Such exploratory analysis, however, is most effective when an abstract model covers the problem space comprehensively with only 3–12 variables. In such cases, the exploration can be comprehensive and comprehensible as a result of recent advances in both theory and technology.[13] In contrast, if one has a large model and explores by holding hundreds of variables constant while varying only a few of them, the results cannot be assessed confidently because it is not known what the effects of varying the others might have been. Many errors of analyses have stemmed from not appreciating the variability and significance of a parameter held constant and taken for granted (e.g., some "authoritative" planning factor that proves very optimistic). In the 1990–1991 Gulf War, for example, the deployment rate of U.S. forces was significantly slower than anticipated in prior

---

[10]Morgan and Henrion (1990), Sterman (2000), Davis (2001a).

[11]Exploratory analysis traces back to the RAND Strategy Assessment System and early ideas for what was then called multiscenario analysis (Davis and Winnefeld, 1983). Many of the concepts matured in the 1985–1993 time period and were reviewed in "Institutionalizing Planning for Adaptiveness," a chapter in Davis (1994). Published applications include Davis and Howe (1988), Davis, Hillestad, and Crawford (1997), Davis (2001a), and Davis , McEver, and Wilson (2002). Colleagues Daniel Fox and Carl Jones have also used exploratory analysis extensively in unpublished work for the Office of the Secretary of Defense, Army, and Air Force. They have used the JICM model (see footnote 16), whereas we have come increasingly to use personal-computer models focused on mission-level analysis rather than theater wars. A mostly independent but closely related stream of research has stemmed from a provocative and influential paper by colleague Steve Bankes (1993), which decried the usual approach to modeling and suggested what would become possible technologically—particularly what he called "exploratory modeling." Bankes and colleague Robert Lempert have subsequently developed related tools and applied them to a number of interesting problems in adaptive strategy, including climate-change policy issues. See, in particular, Bankes (2002) and Lempert (2002).

[12]Another problem is structural uncertainty, which is also called model uncertainty. Its effects can be studied to some extent by parameterizing the structure of the model and exploring the consequences of different coefficient or exponent values, but other methods are typically needed— such as extensive empirical information.

[13]Davis, McEver, and Wilson (2002) push the envelope on this, with some 12 variables.

studies that had assumed different decisions and that had underestimated congestion problems at a few airfields.[14]

A special case of such exploratory analysis is the design challenge. In a myriad of fields ranging from the building of radars and aircraft to the development of systems of systems, top designers need relatively simple models that allow them to consider a broad range of possibilities and tradeoffs before turning the problem over to those who must do detailed engineering.

Good simple models are ubiquitous, but one example is the familiar radar equation. In one of its simpler forms, the formula for the signal-to-noise ratio of a radar receiving the echo from a target is given by

$$S/N = \frac{PG_T G_R \sigma}{R^4}$$

where S is signal strength, N is noise strength, P is the transmitter's power, G_T is the gain of the transmitter, G_R is the gain of the receiver (assumed here to be colocated), σ is the target's effective cross section, and R is the target's distance. The equation is the result of simple physical reasoning and geometry. It can be solved for the distance at which the target is detected, which is given by

$$R_{\text{det}} = c\left(PG_T G_R \sigma\right)^{1/4}$$

where c is a constant (in practice, somewhat of a fudge factor, often determined empirically for a given class of radars and circumstances).

This equation is a highly abstracted version of reality. For example, the target's effective cross section may have little to do with its physical cross section and may depend implicitly on the radar's frequency, the target's shape, and other factors. Further, specialists spend their entire careers designing antennas, which are represented here by nothing more than gain factors. Despite its approximate nature, however, the equation has long been useful. One reason is that for many purposes it can be considered *complete*, although aggregated. That is, if someone asks about a particular detailed variable (e.g., the shape of the target), the response is, "Don't worry, it is not left out; it is reflected through the variable σ, which is the *effective* cross section, not a simple geometric cross section." When conducting an analysis of missile defense, for example, we could vary P, G_T, G_R, and σ and know that we are addressing all of the issues. In contrast, if the

---

[14]See Lund, Berg, and Replogle (1993).

equation left out the target's size (or treated it as fixed by absorbing its value into the constant), then a major factor of the problem would be invisible.[15]

Other Reasons. Such simple models are needed for other reasons as well. They are often much less demanding and expensive to deal with than large and complex models. Run time is a consideration, of course, but other factors are usually more important. Simple models require much less data, data preparation, and post-processing. Analysts can comprehend them and their inputs and outputs quickly. Finally, we note that our knowledge of the world often comes in the form of aggregate information, which in some ways is easiest to interpret with low-resolution models.

The Problem: We Frequently Do Not Have Good Low-Resolution Models

Unfortunately, in many practical problems we do not have any good low-resolution models to start with. Instead, we may have a highly complex model, which lacks the above virtues. Examples in defense work related to effects-based operations include the theater-level models TACWAR, TAC THUNDER, STORM, and JICM.[16] The Department of Defense is currently developing a theater-level model called JWARS. Many of the inputs treated as definitive data in such models are actually quite uncertain. As a result, the number of significantly uncertain inputs is in the hundreds or thousands. Although some of these models were designed with multiresolution features (especially JICM),[17] the models as a whole are inherently large and complex.

Unfortunately, developing simpler models can be difficult. Consider the two obvious approaches:

Deriving a Simpler Model by Deconstructing and Then Simplifying a Detailed Object Model. In principle, it ought to be possible to study that detailed model and develop simpler versions by introducing approximations that simplify equations. In practice, the object models are often difficult to understand analytically. They

---

[15]In some instances, the low-resolution model fits the level of discussion. In others, however, the low-resolution model is merely a second-best surrogate for a detailed model and the problem is to make high-resolution predictions as best one can using a combination of data from the high-resolution model and much more data from the low-resolution model. For such instances, there are good experimental-design methods to improve quality of predictions. See O'Hagan, Kennedy, and Oakley (1999) and Kennedy and O'Hagan (2000).

[16]We omit spelled-out names because the models are known by their acronyms.

[17]Multiresolution features are often claimed but are present only in the sense that detailed models can be asked to generate and display the values of aggregate-level variables. This may be quite helpful, but it does not go far in addressing the reasons for low-resolution models discussed earlier.

may be poorly documented (if documented at all, beyond some instructions on how to change inputs and interpret outputs), may include a variety of ad hoc patches added over a period of years, and—in any case—are almost always in the form of computer programs with no well-specified analytical model (or even a clear conceptual model). And, of course, the quality of the programming itself may be poor or the language used may no longer be in common use. As a result, someone attempting to develop a sound low-resolution model analytically may be forced to do a tedious and lengthy deconstruction, one that may take months if it is feasible at all. This approach *can* be taken sometimes, but it is typically difficult unless the original developers are still available.

Starting from Scratch. A common inclination by an analyst working on problems of strategy and policy is to forget about the "big model" and simply develop from scratch what he sees as a reasonably good low-resolution model. This, however, can be a substantial undertaking when the problems being worked are complicated. In our experience, it has often proven possible for teams to develop initial versions of simple models in a matter of days or weeks, but those models turn out to need embellishment, testing, and data collection. In the end, a number of man-months may be needed to generate a respectable stand-alone model and its documentation. Such an effort is often quite worthwhile, but it is seldom easy. Further, the simple model that results may have no standing within the organization accustomed to using the more detailed object model (and imagining it to be better and more predictive than it actually is). This is not just because of parochialism, but also because of the difficulties in evaluating "simple models" by inspection: without more detailed comparisons with the object model, how can the organization be confident that the simple model is reasonably accurate?

Why Can't We Just Study and Mimic the Behavior of the Object Model? Such difficulties have led many workers over the years to discover statistical metamodeling. That is the subject of the next section.

Metamodels

Definition

A metamodel is a relatively small, simple model intended to mimic the *behavior* of a large complex model, called the *object model*—that is, to reproduce the object model's input-output relationships. Metamodels have different names in the literature, including "response surfaces" and "repro models." Metamodels are generally thought of as statistically inferred constructs, as suggested in Figure

1.2. That is, one starts with an object model, runs that model many times (preferably as guided by an experimental design),[18] collects the resulting data, and infers a statistical model from those data using one or another form of regression. By using stepwise regression, one can discard inputs that show up as statistically insignificant.[19] Often, one can combine some of the inputs and eliminate others that prove redundant. The statistical metamodel, then, will likely have relatively few variables.[20]

A notable point is that a statistical metamodel has nothing to say about physical or behavioral phenomena, or of cause and effect; instead, it purports only to represent the object model's behavior in an input-output sense. The object model is treated like a black box and the metamodel is seen as just a machine that makes predictions. Usually, one never even sees the statistical metamodel written out as an equation because it would be meaningless to the eye.

RAND*MR1570-1.2*

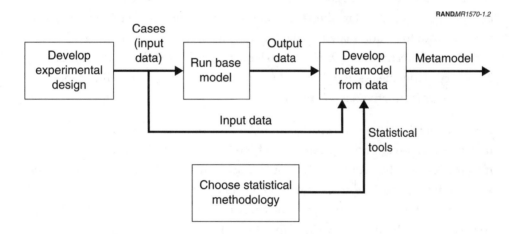

Figure 1.2—A Common Way to Develop a Statistical Metamodel

---

[18]The experimental design should take into account how the model will be used in the application for which the metamodel is being developed. This relates to the more general issue of what Bernard Zeigler calls "experimental frame" (Zeigler, Praenhofer, and Kim, 2000, or Zeigler's chapter in Cloud and Rainey, 1995).

[19]"Significant" in statistics does not mean "important" or "large," as it does in common speech. It means that the result is sufficiently unlikely under the null hypothesis to justify rejection of the null hypothesis in favor of the research hypothesis.

[20]The quality of statistical prediction is intimately associated with experimental design in metamodeling work. Also, there are many subtleties. Stepwise regression, for example, is not always the best approach (Rawlings, 1988), working best when the input variables are uncorrelated (Saltelli et al., 2000). In other cases, the relative "importance" imputed to inputs can change as inputs are added to or deleted from the model.

We regard this as a substantial drawback of a statistical metamodel.[21] To circumvent the problem, one can instead attempt to build a theory-based or phenomenological metamodel. To construct this kind of metamodel, one writes down the functional form based on the physical phenomena represented in the object model, knowledge of the structure of the object model, or a combination. Typically, the function will contain a few undetermined parameters, which must be calibrated so that the metamodel fits the outputs of the object model. Although calibration may involve elementary statistics, the result is not thought of as a statistical model.

Occasionally a phenomenological metamodel will fit the object model to near perfection. When it does not—the more usual case—a combination of the two approaches will often be more successful than either one alone. We have coined the term *motivated metamodel* for a metamodel whose structure is motivated in part by phenomenological considerations, but is also determined in part by data analysis using statistical methods. If we build such a motivated metamodel, it may turn out to be dominated by the theoretical considerations, to have significant corrections and factors, or to look not at all like what was suggested by theory. Nonetheless, if it emerged from a motivated-metamodeling process, we consider it a motivated metamodel.

Figure 1.3 suggests how the results might be assessed. After developing a metamodel, one can run new cases with both the object model and the metamodel, and then compare the results (top of figure). Another way to test the metamodel is to ask questions that are more abstract in nature (e.g., "what would be the benefit of doubling the quantities of resources?"). In this case, the comparison is a bit trickier, because the abstract question may map into a single metamodel case but into multiple object-model cases (e.g., "what would be the benefit of increasing the number of Resource 1 by a factor of 2.5, that of Resource 2 by 1.5, . . .). Nonetheless, such comparisons are unavoidable and can be quite useful.[22] Other ways to validate a metamodel include comparing its predictions to those of some other model of similar resolution, to historical data (which are often aggregated in nature), or to expert opinion. As with validation generally,

---

[21]One use of a metamodel is to explain and justify the behavior of the object model, i.e., to build confidence. Without a simple explanation, one is reduced to arguing that the answers are right "because the (object) model says so." Since a statistical metamodel provides no explanations, it cannot serve this purpose.

[22]The important questions asked of models *often* are higher-level in nature and impossible to map uniquely to the high-resolution model's inputs. To compare models reasonably in such cases, one should test the robustness of conclusions against different assumptions about the ambiguous mappings. If conclusions depend on the mapping, then judgments must be made about what is realistic. For example, if "resources were doubled," would it be more likely that the increases would be spread proportionally; would there be an optimal way to do so that would be realistic, and so on?

RAND*MR1570-1.3*

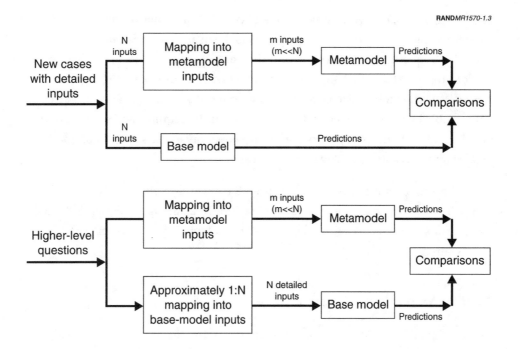

Figure 1.3—Validating and Comparing a Metamodel

there is seldom an easy way to do such things rigorously, except in the classroom with toy problems.[23]

Why Metamodels May Actually Be Rather Good

Why would we imagine that a metamodel could substitute adequately for a more complicated object model? The first reason is that large and complex models often contain many variables that could have been omitted if there had been an effort to introduce approximations along the way. However, developers do not introduce those approximations because they want to maintain flexibility (who knows what variable a particular user may be interested in?) or because they are uncertain for what circumstances the model will be used. In a given application, however, those approximations could be made—if only one could understand the relevant mathematics and coding. Unfortunately, the model is often opaque to users.

A second reason is that the uncertainty in some variables turns out to be numerically unimportant because various effects "average out" (or, at least,

---

[23]For discussion of verification and validation, see, for example, Sargent (1996); Kleijnen (1999); Law and Kelton (1991), Chapter 12; Box and Draper (1987); Balci (1994); Pace (1998); and Wright (2001). Easterling (1999) describes current research in statistical analysis of computer experiments.

average to a constant) over quite a range of cases. That, however, may be altogether unclear to the developer as well as the user. They may believe that the variables in question are important and they may have gone to a great deal of work to represent them. A third reason is that large, complex models often contain a great many variables for no reason other than the insistence of those who established "requirements" for the model. If the requirement setters are members of a committee of users, especially nonanalyst users, the likelihood is high that unnecessary variables will be present.

For these and other reasons, then, it is often the case that a model with N inputs will behave, with reasonable values of those inputs, as though it depended only on m inputs, with m<<N. If so, then a metamodel with m inputs—the right m inputs—may be quite useful.

Typical Methods of Metamodeling

The considerable literature on metamodeling is not reviewed here.[24] Instead, we merely assert that the typical approach involves linear regression or a generalized version of linear regression in which the candidate variables of the regression are allowed to be quadratic or, occasionally, more complex combinations of the elementary variables. That is, if one has a set of inputs X_1 and X_2 and a set of outputs Y from the object model (or from empirical measurements of some type), then the regressions may be:

- Linear (e.g., $Y_{est} = C_0 + C_1X_1 + C_2X_2$)
- Quadratic (e.g., $Y_{est} = C_0 + C_1X_1 + C_2X_2 + C_3X_1^2 + C_4X_2^2 + C_5X_1X_2$)

Even standard statistical packages make it easy to develop the corresponding regressions. Some researchers may use considerably more sophisticated methods (higher-polynomial fits, cluster methods, and so on), but the idea is the same.[25]

As mentioned above, it is also common for metamodels to be developed that treat the object model as a "black box." That is, the person applying the statistics does not know and does not even want to know the innards of the object model or the theory that underlies it. He is merely applying statistical methods to a set

---

[24]See, for example, Law and Kelton (1991), Box and Draper (1987), and Fishwick (1995).

[25]Statisticians argue that starting with higher-order polynomials and simplifying them via stepwise regression is preferred over adopting the linear model at the outset. Parameter explosion can be a problem, which is one reason that "nonparametric" methods, such as smoothing splines or Kriging, are used. These can detect nonlinearities and interactions without a massive increase in parameters. Nonstatisticians, however, often prefer to start simply and add complexity as necessary.

of data.[26] From the viewpoint of a computer scientist, statistical metamodeling is attractive because programs can be built to accomplish such metamodeling more or less automatically. Anything "automatic" is also attractive to cost-cutters looking for ways to reduce the number of analysts required in their organization.

Concerns About Metamodeling

Even simple statistical metamodeling can be very useful. However, we undertook this research because we were skeptical about the routinized approach described above when the metamodel is intended to support analysis of policy or strategy problems. Our concerns were

- The usual metamodels are mathematical constructs, often with little if any intuitive value to decisionmakers.

- As with other aggregations, metamodels are created by accomplishing certain averages. We wanted to understand better what was lost by doing so. Related to this, we have found over the years that merely because a statistical model has what is usually thought of as a good R^2 does not necessarily mean a great deal. This common statistical measure of goodness of fit is hard to interpret (see Appendix D).

- We were skeptical about the metamodels' suitability for work in analyzing systems in which mathematically nonobvious nonlinearities play an important role.

- We noted that some of the statistical packages allow users to calculate the putative significance or "importance" of variables. We were skeptical, however, about how such statistical "importances" relate to the importances seen by a commander in battle or a senior official or general officer engaged in peacetime resource allocation. Would the "importances" generated by statistical packages be approximately right for such purposes, or misleading?

- Finally, because of our personal interdisciplinary inclinations, we found pure statistical metamodeling to be distasteful, given that much is known about the real-world systems being described. Why should we not be using some of that information? And why shouldn't explanations be in causal terms if at all possible?

Other researchers with the same concerns would argue for improving the sophistication of statistical metamodeling, and much is probably possible by

---

[26]For a brief tutorial on this type of work, see Kelton (1999).

14

doing so,[27] but our interest is largely in models that are understandable in the natural language of applications, rather than complex mathematics.

Approach

With this background, then, our approach was to investigate how metamodels can be improved by combining the virtues of statistical methods and theory-informed methods. Figure 1.4 suggests the basic concept, as a contrast to Figure 1.2. The key point is that our intention was to draw upon an understanding of relevant cause-effect phenomena to suggest the analytical *form* of the metamodel. The next chapter describes our approach in more detail.

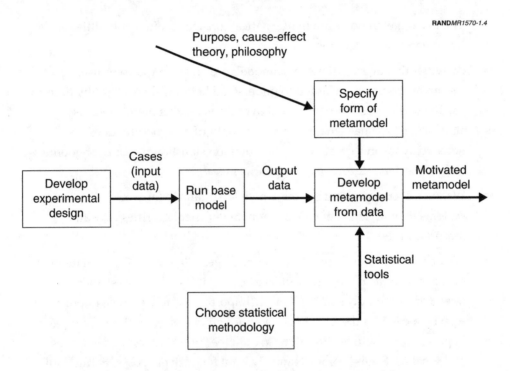

RAND*MR1570-1.4*

Figure 1.4—Improved Development of a Metamodel

---

[27] A number of advanced methods exist. For an introduction to the global sensitivity analysis literature, see Saltelli et al. (2000). Quasi-regression (An and Owen, 2001) can provide an inexpensive prediction method that also identifies anomalies. "Kriging" (named after an individual) models response surfaces at the local scale through correlation structure, rather than more globally through mean structure. Kriging is a more powerful technique than regression when nonlinearities and interactions occur (Sacks, Welch, Mitchell, and Wynn, 1989; Handock and Stein, 1993; and Cressie, 1993). We thank colleague Brian Williams for pointing us toward this literature.

2. Hypotheses, Experimentation, and Iteration

Hypotheses

We began our work with extensive discussion and hypothesis sketching. Our hypotheses, which drove the design of experiments, were strongly influenced by our initial skepticisms about statistical metamodeling, as discussed in the previous chapter. The hypotheses were as follows:

- *Hypothesis 1*: With only modest-to-moderate investment of time, it will often be possible to develop a reasonable, albeit much simplified, understanding of the problem being analyzed, and to translate that understanding into useful guidance for metamodeling.

- *Hypothesis 2*: I t will often be particularly important in such work to worry about "critical components" of the system under study. That is, if the analysis involves assessment of a system (e.g., the combination of capabilities needed to suppress air defenses or interdict and stop maneuver forces), it will be fruitful to view the problem from the perspective of ensuring that *all* of the critical components are present with enough numbers and effectiveness. This will lead to distinctly nonlinear forms for the proposed metamodels (i.e., products of variables, rather than sums).

- *Hypothesis 3*: It will often be possible to generate motivated metamodels that will be understandable and useful to decisionmakers. Although decisionmakers generally will not be interested in analytical details or methods, some will value simplified analytical descriptions such as transparent one-line formulas with understandable factors or terms.

- *Hypothesis 4*: The "importances" generated by routine use of packages and statistical metamodels will often prove to be highly misleading in dealing with system problems involving critical components or a malevolent adversary who can determine circumstances of conflict.

It was not possible, within a small research project, to do justice to these hypotheses. In particular, we could do nothing rigorous about assessing whether the word "often" is valid in the hypotheses. Nonetheless, we believed that hands-on experimentation would sharpen our understanding of the issues and, in the process, cause us to become more or less bullish in our claims. Further, we believed that going through actual calculations would be useful because it is common in analytical work for notions to be valid in some theoretical sense, but ultimately not very important quantitatively or qualitatively. As we noted in the previous chapter, metamodeling often proves much more accurate than might be expected a priori for all sorts of reasons related to the averaging out of conflicting effects, the small numerical size of some factors, and so on.

An Overview of the Experimentation

We conceived an experiment to explore these hypotheses and observe issues in more detail. Instead of working directly with any of the very large and complex legacy models that require a great deal of effort, we would use a relatively small but complex model with which we were familiar and that could be run conveniently on a personal computer to generate as many cases as we could possibly use. As discussed in Appendix B, we used a model with scores of variables and many discontinuities and other nonlinearities. It had taken a number of months to build. Although it was seen as a "simple model" in its original context (when used in preference to a full-up theater-level combat model), it seemed more than adequately complex as a surrogate for a complex object model in our research.

Using our object model, we would then generate data and develop a series of metamodels. The first would be a garden-variety statistical metamodel requiring essentially no understanding of the object model other than its list of inputs and outputs.[1] We would then draw on increasingly rich expressions of theory to motivate subsequent metamodels. We would observe ourselves as we built them, notice the kinds of issues and choices that arose, and assess the various metamodels produced. Our intention, then, was to understand better what was involved in "motivating" metamodels and how much value was added by different levels of theory. Although the detailed comparisons would obviously be determined by the particular problem area being modeled, and the nature of the object model, we hoped to find insights with relatively broad applications. Our experiment, then, was a matter of discovery and exploration, rather than of rigorous testing.

---

[1] A methodological dilemma for us was whether to have the baseline statistical metamodel represent "normal" practice or what might be built by a first-class statistician with the time to use more advanced methods. We chose the former because we believed that this would be more relevant and understandable to the target audience.

Criteria: What Makes a Metamodel Good?

Because we were to be comparing and assessing a series of metamodels, we needed to establish some criteria for doing so. We settled on the following criteria, although what follows has been "neatened up" in the course of our work:

Goodness of fit. Obviously, we want a metamodel's predictions to be reasonably consistent with those of the baseline model. A straightforward measure of this is the root mean square error (or fractional error) of predictions across the relevant domain of input values. For our purposes, this is superior to the commonly used R^2 (see Appendix D).

Parsimony. For both cognition and exploratory analysis, a good metamodel will have relatively few independent variables. Parsimony may be achieved by omitting some of the baseline model's inputs (i.e., treating them as constant, after concluding from stepwise regression that we could do so) or by combining several object-model inputs into a smaller number of intermediate variables. The set of independent variables should be rich enough to represent the issues being addressed with the model. Beyond that, the fewer extra variables, the better.[2]

Identification of "critical components." Our third criterion seems new and we believe it to be crucial.[3] Many uses of models in analysis involve systems or strategies, the failure of which is to be very much avoided. We suggest that a metamodel should highlight all of the input variables that are essential to success—especially when troublesome values of those variables are plausible. That is, appropriate variables should be created and even "forced" into the model's structure because one knows that they represent critical components. When viewed analytically, the model should not give the impression that one can compensate for a weak component of the system by improving some other component (if such substitution is in fact inadequate). This is a significant consideration in metamodeling, because standard statistical methods consider linear models that imply substitutability. We refer to components that must individually succeed simultaneously (have values above or below an appropriate

---

[2]Occam's razor ("keep it simple") is sometimes thought of as folk wisdom, but it can be motivated by analytical arguments. There are measurable dangers in "overfitting" available data with too many unknowns; the result is a better fit to the data, but, quite often, a poorer fit to new data. Occam's razor can also be discussed in terms of Bayesian theory. For discussion by Jacob Eliosoff and Ernesto Posse, see the interactive web site at http://cgm.cs.mcgill.ca/~soss/cs644/projects/jacob/applet.html.

[3]Few ideas are truly new and this may be no exception, but we have not observed the point being made elsewhere. See Davis (2001b) and Davis (2002b) for defense-planning studies that emphasize the critical-component issue.

threshold) as *critical* components. If critical components in this sense exist, the metamodel should be appropriately nonlinear.

Story line. Without a story, a model is just a "black box." A story explains why the model behaves as it does. Moreover, it relates the model to the real world, telling us why the model *should* behave as it does. We use the term "story line" because all models are simplifications of reality, but we intend no cynicism. Said differently, the model should be physically (or psychologically) meaningful and interpretable. Ideally, it should also be describable by one or a very few simple and transparent formulas that convey readily the key factors and how they affect the problem.

Even analysts using a very large model find it necessary to invent stories to explain its behavior. When the story is more or less concocted, without being able to infer it readily in the model's structure, the explanation may be rather anthropomorphic as in, "Well, the model knows that . . . and so it . . . " Although that may not be satisfying, it is better than saying, "Well, the model says . . . ." Our point, however, is that the need for a story line is well known to anyone who has briefed a client who has questions.

Some stories are much better than others. The story line should not only be reasonably accurate, it should be as general as possible, rather than merely explaining results for base cases and minor variations. Describing the behavior of falling bodies is better done with a story that refers to both gravity and drag than with a story that mentions only gravity, and therefore applies only in the upper atmosphere.[4]

Characterizing importance of variables. Toward the end of our work, we added this criterion, because we discovered that it was more of an issue than we had previously recognized.[5]

With this background, we now describe the analytical experiments we conducted to illustrate and sharpen our understanding of issues and methods. Our focus is less on the experiments per se (but see Appendix B) than on what we learned along the way and the inferences we drew.

---

[4]The story, of course, is likely to assert causality and one must then be cautious. There should be no hidden variables correlated with any of the variables maintained in the metamodel. Recall the old saw: "I had a whiskey and soda, a bourbon and soda, and a gin and soda; they all made me woozy, so it must be the soda." A partial protection here is to add to the metamodel additional variables to assess the effect of hidden (omitted) variables. If their coefficients are large, then problems exist.

[5]See Saltelli et al. (2000) for discussion of additional criteria related (for example) to global sensitivity analysis, including Fourier Amplitude Sensitivity Test (FAST) or the method of Sobol.

3. Results of Experimentation

Defining a Baseline for Comparison: Pure "Statistical Metamodels"

The Approach Taken

One of the more troublesome aspects of our experimentation was deciding how to go about "pure" statistical metamodeling, so as to have a baseline against which to measure the goodness of "motivated metamodels." Any senior analyst building a metamodel has his own bag of tricks. Further, statistical methodology continues to improve (see earlier citations). Thus, it is a bit misleading to refer to "standard" or "normal" metamodeling as though it were cut-and-dried. As mentioned earlier, however, we chose a baseline of multiple linear regression— something available to everyone who owns Microsoft EXCEL, let alone more specialized tools such as SAS or STATA.

To explain the issues here, let us describe a toy problem, one not dissimilar to the actual military problem treated by our model (Appendix B). Suppose that the problem is simply accomplishing a certain amount of work W (e.g., tons dug, lines of software written). Suppose that the resources for accomplishing the work builds with time, starting at R_0 and increasing at rate B(t) (additional resources added per day) until the job is completed. Suppose that the productivity of resources is P(t): that is, an amount of work P(t) will be accomplished each day by each unit resource, its value increasing from P_0 to P_f in some complicated way characterized by an equation with parameters f, g, and h. Perhaps B(t) is specified by B_0, the initial buildup rate valid for time d and a subsequent buildup rate B_1 valid thereafter.

We may have a black-box computer model that predicts the time T to accomplish the job. It is a computer program, the details of which have been forgotten, but it generates T, based on inputs W, R_0, B_0, B_1, P_0, P_f, d, f, g, and h.

RAND*MR1570-3.1*

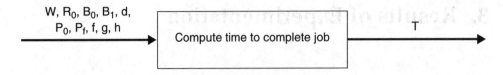

Figure 3.1—A Black-Box Model Computing Time to Do Job

A statistical metamodel to emulate the black-box model might be a simple linear regression, where the coefficients $C_0, C_1, \ldots$ would be determined by fitting the formula to a sampling of data from the black-box model.

$$T = C_0 + C_1 W + C_2 R_0 + C_3 B_0 + C_4 B_1 + C_5 d + C_6 P_0 + C_7 P_f + C_8 f + C_9 g + C_{10} h \qquad (3.1)$$

If this model did not fit very well, as judged by the R^2 criterion or the root mean square error, then the analyst might try something more complex. He might, for example, consider quadratic factors of the inputs, in which case he would have

$$T = C_0 + C_1 W + C_2 R_0 + C_3 B_0 + C_4 B_1 + C_5 d + \ldots + C_{10} h + C_{11} W^2 + \ldots$$
$$+ C_{20} h^2 + C_{21} W R_0 + \ldots \qquad (3.2)$$

He would still find the coefficients by using the methods of multiple linear regression, but the regression variables would now include some composite variables such as WR_0, which are quadratic with respect to the original input variables. This statistical metamodel, then, is linear in one sense but quadratic in another. Still, no "understanding" of the model has been required.

In principle, the metamodeler could consider cubic terms or even higher-order polynomials, although that is relatively unusual outside of academic research. It is even more unusual to automatically consider other forms, such as regression variables that are exponential in the various elementary inputs, such as e^{-CW}.

In our work, we decided to treat the linear regression model as the baseline statistical metamodel against which we would compare motivated metamodels. This is by far the most commonly reported approach to metamodeling, although we recognize, of course, that any given statistical metamodeler might use more advanced methods and some personalized tricks, which might or might not reflect subject-area knowledge or knowledge of the object model's innards.[1]

---

[1]Clustering techniques, for example, might be used to partition the input space into regions, and different metamodels fit in each region. For a survey of clustering techniques, see Treshanskyand McGraw (2001).

A Path Not Taken

Interrupting our flow briefly, we note something that we did *not* do in our limited observational experiments. In particular, we did not develop baseline metamodels reflecting state-of-the-art practice by professional statisticans (e.g., practice using techniques cited throughout the paper). Instead, our baseline metamodels were deliberately simple, calling upon only methods and tools that are in common use in the analytical community. These, we believed, were the appropriate baseline for our target audience. Further, our primary interest is in models that are understandable in the natural causal-language terms of the application area, rather than the sometimes arcane language of advanced mathematics. Taking this approach, however, made it impossible to draw conclusions about how much value adding phenomenological information would have when compared to a more sophisticated baseline of statistical metamodeling. Further research is needed on such matters and, more generally, on how best to use improved statistical methods, improved personal-computer-level tools, and phenomenological knowledge together.

Having posted this caveat, let us now return to the flow of what we did.

Types of Theoretical Knowledge

General Observations

If subsequent metamodels would incorporate increasing amounts of problem-specific knowledge, what kinds of knowledge might we bring to bear—in our specific problem and more generally? One consequence of our experimentation was a great deal of discussion between us. We had the following observations:

1. A sensible approach to metamodeling might include trying composite variables constructed to have consistent dimension. For example, if the output being calculated is a distance, it is reasonable to construct composite variables with the dimensions of distance. Thus, if inputs include various times and an average speed, then the times multiplied by an average speed would be candidate regression variables. This would require very little "theory" (and might even be wrong-headed), but it goes a bit beyond normal metamodeling and in some cases improves results.

2. From even minimal model documentation, it may be possible to identify intermediate variables that are composites of the complex model's inputs. Such intermediate variables may be identified for generating explanatory displays or supporting analysis, even if the complex model was not itself

designed with multiple levels of resolution. Such intermediate variables, which may be quite nonlinear, are good candidates for regression variables.

3. Even without such minimal documentation, it may not require extraordinary problem-specific physical reasoning to guess appropriate intermediate variables to try. For example, in the toy problem above, it takes no genius to think that an equation predicting time to do work would have natural terms such as W/P_0R_0 (even if based only on dimensional analysis—i.e., on ensuring consistent units from term to term).

4. From even minimal reading of model documentation or discussion with a base-model expert, it may be possible to identify what amounts to alternative branches to the simulation. For example, the simulation may find the better solution of two or three candidates and report that. If so, then the quality of metamodels will likely be improved by building separate metamodels for the several cases and linking them by something like a "Take the best solution" algorithm. Alternatively, the branches could be incorporated in a single regression model by means of indicator variables. We show an example of this at the end of this list.

5. By focusing on the output of most analytic interest, and by then using a combination of simple physical reasoning and heroic assumptions about the validity of using averages for various quantities, one may be able to estimate the analytical form in a dimensionally correct way (see example below).

6. If, in addition, one looks for individually critical components and then makes an effort to reflect this phenomenon with a *product* of factors, one can build in extremely important and nontrivial nonlinearities, albeit as an approximation of unknown accuracy. That is, if a system would fail if any of three capabilities X, Y, or Z were zero, then we should expect that system effectiveness would go as

$$E = C_1 XYZ[1 + C_2 F(X, Y, Z)]$$

where C_1 and C_2 are constants and F is some function of X, Y, Z, perhaps a linear sum (see example below). Appendix B gives examples.[2]

7. With some modest algebra and calculus, one may be able to estimate analytical solutions that take into account, for example, straightforward buildups of resources.

---

[2]The machinery for including such interaction terms in the metamodel already exists in standard statistical packages such as SAS.

Illustrating the Ideas in 4, 5, and 7

Returning to the toy problem described above, suppose first that we think about it mathematically and physically without trying to derive an exact solution. In fact, let us be deliberately crude. We will assume, as a start, that the buildup of resources proceeds linearly at the initial rate B^0. Further, we replace P(t) by an average, estimated roughly as just $(1/2)(P_0 + P_f)$.

With this approximation, we can solve the remaining problem with nothing more profound than would be understood by a student of undergraduate calculus. To a first approximation, the solution for T is given by a simple integral equation:

$$P \approx \frac{1}{2}(P_0 + P_f) \tag{3.3}$$

$$\frac{W}{P} \approx \int_0^T \{R_0 + B_0 s\} ds = R_0 T + \frac{1}{2} B_0 T^2 \tag{3.4}$$

Rearranging and using the quadratic equation of elementary algebra, we obtain

$$T \approx \frac{-R_0 + \sqrt{R_0^2 + \frac{2B_0 W}{P}}}{B_0} = \frac{R_0}{B_0}\left\{\sqrt{1 + \frac{2B_0 W}{R_0^2 P}} - 1\right\} \tag{3.5}$$

Despite the crudity, the approximation tells us a lot. In particular, we now expect the true expression for T to depend on something roughly like the right-0hand side of Eq. (3.5), a highly nonlinear composite of elementary variables. This, then, would be a good regression variable to try. We might try

$$T = C_0 + C_x \frac{R_0}{B_0}\left\{\sqrt{1 + \frac{2B_0 W}{R_0^2 P}} - 1\right\} + C_1 W + C_2 R_0 + C_3 B_0 + C_4 B_1 + C_5 d$$

$$+ \dots C_{10} h + C_{11} W^2 + \dots C_{20} h^2 + C_{21} W R_0 + \dots \tag{3.6}$$

We might find that the bracketed quantity dominates results of regression, with other terms adding up to a small net correction. In fact, we could test that from the outset by replacing the long set of terms with a constant and seeing how large the net error turns out to be. If it proves small, we could just skip the complexity of the more extensive regression (although we would be forgoing any opportunity to make predictions about the effects of, for example, d, f, g, and h).

The next step might be to consider explicitly the time dependence of buildup B(t). Being more careful now, we would discover that there are two possible integral equations, depending on whether the time T turns out to be greater than the delay time d. We would define T_1 and T_2 by

$$\frac{W}{P} = \int_0^d \{R_0 + B_0 s\} ds + \int_d^{T_1} \{R_0 + B_1 s\} ds$$

$$\frac{W}{P} = \int_0^{T_2} \{R_0 + B_0 s\} ds \tag{3.7}$$

Then $T = T_2$ if $T_2 < d$ and $T = T_1$ otherwise.

We can solve these by integrating, rearranging, and using the quadratic equation:

$$\frac{W}{P} = R_0 d + \frac{1}{2} B_0 d^2 + R_0 T_1 - R_0 d + \frac{1}{2} B_1 T_1^2 - \frac{1}{2} B_1 d^2$$

$$T_1 = \frac{R_0}{B_1} \left\{ \sqrt{1 + \left[\frac{2B_1}{R_0^2}\right]\left[\frac{W}{P} + \frac{1}{2}(B_1 - B_0) d^2\right]} - 1 \right\} \tag{3.8}$$

$$\frac{W}{P} = R_0 T_2 + \frac{1}{2} B_0 T_2^2$$

$$T_2 = \frac{R_0}{B_0} \left\{ \sqrt{1 + \frac{2B_0 W}{R_0^2 P}} - 1 \right\} \tag{3.9}$$

Then $T = T_2$ if $T_2 < d$ and $T = T_1$ otherwise. This only somewhat more sophisticated exercise in mathematics suggests additional candidates for the metamodel. The metamodel might be assumed to be

$$T = Min[T_1, T_2]$$

$$T_1 = C_1 \frac{R_0}{B_1} \left\{ \sqrt{1 + \left[\frac{2B_1}{R_0^2}\right]\left[\frac{W}{P} + \frac{1}{2}(B_1 - B_0) d^2\right]} - 1 \right\} + C_2$$

$$T_2 = C_3 \frac{R_0}{B_0} \left\{ \sqrt{1 + \frac{2B_0 W}{R_0^2 P}} - 1 \right\} + C_4 \tag{3.10}$$

If the values of either C_2 or C_4 turn out to be large, we can go back and try the other regression variables such as in Eq. (3.5). If not, we would note the small errors and proceed without further detail.

Finally, of course, we could try to understand better the expression for P(t) and see about developing an analytical solution. Let us assume, however, that doing so would be too much trouble—we do not really understand the object model in this regard. Thus, we would stop trying to add "theory" to inform the metamodel at this point.

What is significant here is that the natural variables to use in a regression are nonlinear composite variables. It is not unreasonable to expect that using those, rather than just linear and quadratic combinations, will yield a much better metamodel (especially if the physical insight of approximating average productivity as the simple average of initial and maximum values is not grossly wrong).

Finally, a word about critical components. In a sense, this effect appears in the toy problem. Note that factors such as W/RP arise. Getting the job done in a short period of time will be impossible if the work is excessive *or* if the resources available are too small *or* if their productivity is too small. Further, although one could compensate for low productivity by having more resources, the substitution would be in multiplicative, not linear terms. That is, if R were only half as large as it needed to be, one could compensate by doubling P. This is not a linear substitution, such as might be predicted from an uninformed regression model.

In other problems, the critical-component issue is clearer. See Appendix B.

How Exploiting Increasing Amounts of Knowledge Improves Metamodels

With the caveat that the results in what follows stem from a specific set of experiments with a specific object model (Appendix B) and only garden-variety statistical methods as a base, it is nonetheless interesting to see how adding information improves results of metamodels. Table 3.1 does so in rather generic terms intended to suggest what may be fairly general insights, even though they came from working the specific problem in Appendix B.

We had originally thought that identifying good "aggregation fragments" (as in rows 2 and 3 in the table would prove quite useful, but—in this case at least— their value was only modest. What made the biggest difference (Metamodel 3 versus Metamodel 2 in the table) was in postulating an analytical form for the solution (much as in the toy problem above), even if doing so required some heroics in terms of replacing conceptual integrals with products of average values of integrands, ignoring various subtleties, and so on. This required form included explicitly the postulated critical-component phenomenon, accomplished simply by assuming that the solution to the problem would vary with the *product* of the separate components of capability needed. As the table indicates, the resulting metamodel did well in all respects.

26

Table 3.1

Value of Theoretical Knowledge Added

Meta-model	Type of Knowledge	Parsimony: Number of Inputs[a]	Average Accuracy	Quality of Story/ Analytical Transparency	Prediction of Critical-Component Phenomenon	Predictions of Relative Importance of Variables
1	Nothing but inputs of object model	14	Poor	Very poor/ very poor	Very poor	Poor
2	+Composite variables (aggregation fragments)	10	Fair	Very poor/ very poor	Very poor (sometimes fair)	Poor
3	+Top-down reductionist structure for form[b]	5	Good	Very good/ very good	Very good	Good
4	+Improved structure based on simple algebra and calculus	5	Very good	Very good/ fair	Very good	Very good

[a]Inputs remaining after stepwise regression eliminates variables with significance less than 0.05.
[b]See Eqs. (3.5)–(3.6) above, for example.

In the experiment, we also noted that the quality of that postulated analytical form could be improved by merely doing some math that would not stress a freshman student of calculus or calculus-based physics (but might be more than a hurried engineer or a mathematically rusty analyst would want to attempt). This generated Metamodel 4, which proved astoundingly accurate. In retrospect, then, the behavior of the object model was not nearly as complicated as one might expect from its description, documentation, and interface when being used. However, the price paid for introducing this additional fillip of sophistication was a loss of analytical transparency. The story could still be told convincingly, but more skill would be needed in the construction of Vu-graphs to make it understandable, and the possibility of pointing at a one-line equation with self-evident logic was sacrificed.

The astute reader will have recognized that many other metamodels might have been considered, with knowledge being added in different orders, and so on. Also, statisticians would in some instances use advanced methods or try special tricks with which they have become familiar (e.g., item 1 in the list of knowledge types). Thus, Table 3.1 is only one possible listing of metamodels and their results. In our work, we considered a number of other metamodels (although nothing very advanced). One interesting result was that with some combinations of more or less purely statistical tricks (nothing more than, say, using dimensional analysis), the metamodel could actually be quite accurate. It is for this reason that we emphasize that *the primary value of adding the theoretical knowledge was to improve story, transparency, treatment of critical components, and the ability to compare the importance of different variables. Those improvements, in turn, were due largely to the step of developing super-simple, super-approximate "formulas" with structures that built in the critical-component phenomenon as best we understood it.* We suspect that this result is more generic than our particular experiment.

4. Summary and Lessons Learned

We began our work by developing a number of hypotheses. Conclusions in response to those hypotheses were as follows.

- *Hypothesis 1*: With only modest-to-moderate investment of time, it will often be possible to develop a "reasonable," albeit much simplified, understanding of the problem being analyzed, and to translate that understanding into useful guidance for metamodeling.

  Conclusion: In the cases we have worked, this proved to be true. The time required might be greater than anticipated (days or weeks, not hours), but it was still much less than the time required to build a new low-resolution model from scratch or deconstruct a complicated computer program and then simplify it.

- *Hypothesis 2*: It will often be particularly important, in such work, to worry about "critical components"of the system being studied. That is, if the analysis involves assessment of a system (e.g., the combination of capabilities needed to suppress air defenses or interdict and stop maneuver forces), it will be fruitful to view the problem from the perspective of ensuring that *all* of the critical components are present with enough numbers and effectiveness. This will lead to distinctly nonlinear forms for the proposed metamodels (i.e., products of variables, rather than sums).

  Conclusion: Our concerns in this regard have been well confirmed in a number of problems that we have studied (e.g., that of Appendix B). The importance of this critical-component perspective has broad implications for analysis generally.[1]

- *Hypothesis 3*: It will often be possible to generate motivated metamodels that will be understandable and useful to decisionmakers. Such decisionmakers generally will not be very interested in analytical details or methods, but some will value simplified analytical descriptions such as transparent one-line formulas with understandable factors or terms.

---

[1]In Davis (2002b), this is referred to as mission-system analysis, by which is meant that the analysis should be framed so as to understand and highlight all of the individually critical capabilities required for a mission.

Conclusion: What is understandable and useful varies enormously with the decisionmaker. Empirical work is needed here, but the hypothesis seems to us sound as stated. In the particular case studied here (Appendix B), we were able to give what we believe were simple, understandable, and useful explanations.

- Hypothesis 4: The "importances" generated by routine use of statistical packages and statistical metamodels will often prove to be highly misleading in system problems involving critical components or a malevolent adversary who can determine circumstances of conflict.

  Conclusion: Phenomenological concepts such as "causation" and "importance" are not necessarily captured in the statistician's ideas of "correlation" and "significance." We identified specific cases (see Appendix B) where standard statistical methods concluded that certain variables could be dropped from a metamodel even though they entered the problem in precisely the same way as variables that were retained. More work is needed to understand how and when such matters arise and how they can be avoided. It is not that the statistical analysis was wrong, but rather that the resulting metamodel could not be used properly for certain resource allocation decisions.

Shortcomings of Our Statistical Metamodels and Conclusions from the Experimentation

As anticipated, the "pure" statistical metamodels that we created had serious problems. These included:

Failure to Tell a Story

As recognized at the outset, a fundamental problem with statistical metamodels is that they do not convey insight (except insights of the form that something appears to have relatively linear, quadratic, or exponential behavior). That is, they do not allow users to "understand," communicate, and discuss issues in a satisfactory way. This is a serious problem in higher-level decision support where decisionmakers must deal with considerable uncertainty and difficult tradeoffs, are loathe to just accept some model's results, and must convincingly explain to others the virtues of their choices.[2]

Critical Components

The second shortcoming is more subtle. If the real-world system being modeled will fail if *any* of several components fail—a common feature of many important devices or operations—this fragility of the system may not be captured by a statistical metamodel—especially when the critical components are intermediate-level abstractions rather than something directly visible in inputs. This is common with bottom-up models, such as military or transportation-system simulations. Further, the issue of critical components may be less a feature of the object model than a phenomenon of particular questions asked of the model. Thus, a statistical metamodel may be fairly accurate broadly, but poor for some questions.

Implications for Resource Allocation

An interesting shortcoming that we had not fully anticipated is that a statistical metamodel obtained with stepwise regression may characterize one input as much more important than another, even though their real importance is identical. Which variables are deemed important is determined as much by the experimental design—i.e., which data points are collected from the object model—as by the structure of the object model. If the object model is $F(X_1, X_2, \ldots X_N)$ and one determines that a metamodel fits the behavior well with only three variables, as in $G(X_1, X_2, X_3)$, it may be that one of the omitted variables, say X_4, appears in F in identically the same way as one of the included variables, say X_3 (e.g., it might appear only as part of the product $X_3 X_4$). This may occur if, for example, the percentage of variability in X_4 in the sample is much smaller than the percentage of variability in X_3. It follows that if the statistical metamodel is used to characterize the relative importance of its inputs with an eye toward resource-allocation priorities, the results may be grossly misleading.[3] This

---

[2]Modern work in statistical methodology aspires to make inferring a story possible, even without preexisting phenomenological insight (see Saltelli et al., 2000, for examples). Success varies a good deal in practice. Based on what we have seen so far (including a shallow look at the advanced literature cited earlier), we remain skeptical about such techniques providing—*by themselves*—a comprehensible story useful to decisionmakers (who often dislike statistical "explanations"), good treatment of nonlinearities such as the critical-component issue, or the highlighting of "corners" that adversaries can exploit. For related discussion of limitations, see Saltelli et al. (2000), pp. 45–46. Perhaps we are wrong in this, of course, but only further research will tell the tale.

[3]A reviewer argues that the problem might not arise with more sophisticated statistical metamodeling using global sensitivity analysis techniques.

reminds us that a variable's "importance" to a statistician is different from that variable's importance to a decisionmaker.

Shortcomings in the Presence of an Adversary

Finally, the statistical metamodel may be inaccurate in circumstances where a competitor or military adversary seeks to exploit the system's weaknesses. The relevant inputs may then be unusual and strategically correlated. The metamodel, which is good on average, may be inaccurate in these corners of the input domain. Figure 4.1 suggests the point with a cartoon in which Z (the vertical dimension of the figure) is a function of X and Y. For most of the input space, Z is approximately 10. However, in one corner, where both X and Y are small, Z is also very small. A simple statistical metamodel might correspond to a plane at roughly a Z value of 10, but tilted slightly so as to range from perhaps 9 to 11, thereby compensating to some degree for the peculiar behavior in the small and potentially obscure corner. If the system being represented involved adversary processes, however, the adversary might be able to choose circumstances focused on that corner. It is the job of generals, after all, to find and exploit their enemy's weaknesses. The moral here is that what may appear to be an obscure corner when constructing and testing a statistical metamodel over the full relevant domain might turn out to be critical.[4]

Benefits of Motivated Metamodels

Based on our experiment and more general reasoning before and afterward, we concluded that the synthetic approach of motivated metamodels can help considerably. In particular, we concluded that by using cause-effect insights of a reasonable nature to motivate the form of regression variables, it should often be possible to develop motivated metamodels that

- Tell a credible story useful for reasoning and communication

- Highlight the independent importances of a system's critical components

- Improve the quality of importance estimates and resource-allocation priorities

- In some cases, also improve average accuracy (although that is less important, since more garden-variety statistical metamodels are often fairly accurate and advanced techniques can improve results further).

---

[4]If one can anticipate such behavior, the troublesome region can be oversampled in an experimental design. Such anticipation, however, would likely depend on phenomenology.

32

RAND*MR1570-4.1*

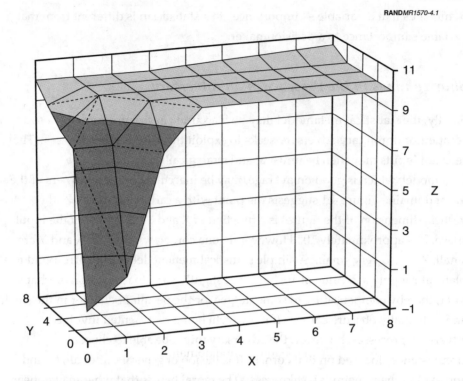

NOTE: A linear metamodel would be a plane surface with values of about 10–11, tipped slightly to reflect data near the origin.

Figure 4.1—A Model with Strange Behavior in an Obscure Corner

A Significant Organizational Benefit of Motivated Metamodeling: Easy Validation

Once built and tested, a motivated metamodel can be described as consistent with the object model in key respects. That, plus its transparency and face-value reasonableness may be all that is needed to obtain the metamodel's quick acceptance by an organization that accepts the object model—except for the proviso that important conclusions of analysis should be spot-checked with the object model. In contrast, gaining organizational acceptance for a new simple model, however compelling to first-rate analysts who have studied the problem, can be extremely difficult.

Suggested Elements of Motivated Metamodeling

Although no general recipe can be provided, some principles seem likely to be broadly useful:

1. *Identify the problem's critical components.* This may require fresh thinking, since the components will often *not* be highlighted in the object model's documentation or be evident in low-level inputs. Finding critical components will require some subject-area expertise or, for example, broad engineering talents, but may not require in-depth mathematics or careful deconstruction of the object model.

2. *Identify significant branches of the object model.* This implies building in a particular form of nonlinearity that corresponds to cases that are sufficiently distinct so as to justify separate metamodels. Doing so will be much easier if the object model's documentation is good.

3. *Identify natural "aggregation fragments" to use as composite variables in statistical analysis* (e.g., one might expect, in a given problem, that variables x, y, and z would ordinarily enter the problem only in the combination xy/z).

4. *Postulate structural forms* by using the three principles above, dimensional analysis, simple physical reasoning, and rough approximations (such as replacing an integral with a representative value of its integrand multiplied by the effective width of the integration domain).

5. *Build in fudge factors.* That is, compensate for imperfections of the postulated forms by building in unknown coefficients and error terms. This would be inappropriate if one were determined to prove the adequacy of a simple model with only a minimum of adjustable parameters, but in problems that are genuinely more complicated, building in such corrections is desirable—especially if one wants to claim calibration with the object model. In practice, it is straightforward to explain, "As you can see from the form of the equation, what is going on is basically . . . . There are complications, however, which affect things on the margin. Those are accounted for approximately by . . . and . . . , which you can think of as small numerical fudge factors to fit behavior of the complex model." Such an explanation, of course, would be unconvincing if the "corrections" proved large.

6. *Where applicable, game the problem being modeled to identify domains of special interest.* These would include worst-case adversary strategies and variables that are strategically correlated. Such gaming will help determine how tests of the object model should be specified in an experimental design or how preexisting output data should be sampled.

Implications for Model Validation and Documentation

A side benefit of our research was recognizing requirements for model validation and documentation that we have not seen emphasized previously. We recommend to organizations such as the Defense Modeling and Simulation Office that

- Future guidance on model validation should highlight the need to test whether the model correctly represents critical-component and adversary-process issues.
- Guidelines for documenting models should admonish authors to identify critical components and adversary-process issues and to consider suggesting approximate analytical forms, either as an aid to understanding or as the basis for possible motivated metamodeling.

Today, even good model documentation seldom addresses these issues unless they are front and center from the outset of modeling (e.g., as in nuclear-reactor safety). Ordinarily, documentation lays out model structure from the bottom up, provides definitions for inputs and outputs, and specifies key relationships. Nor does documentation routinely offer up approximate analytical relationships that might be useful in simple-minded reasoning or motivated metamodeling. Such information *may* be provided, because the author "thinks that way," but not because it is expected. This is a pity, because developers of documentation are often in an excellent position to provide the related information. Those who pick up the model later may have a lengthy learning curve before they are able to make the same observations (if, indeed, they even aspire to such a deep understanding of the model).

Possible Resistance to Motivated Metamodels

It is ironic that the approach we describe needs to be discussed. In a sense, it is obvious and some of its aspects are well precedented in science and engineering, where researchers often see experimental data merely as a way to calibrate a theory. However, we have observed in the analytical community a rather sharp disciplinary divide into three groups. One group routinely uses statistical methods and an empirical approach in which the object model is treated as a black box. The form of the resulting models is driven by the on-the-shelf methods of statisticians, not physical insight into the system being represented. That is regarded as advantageous in many respects, because it corresponds to "allowing the data to speak," rather than biasing results with one or another theory. The second group routinely uses detailed cause-effect models, often in the form of simulations. These can explain results well to other aficionados, but the explanations may be complex and unsatisfying for higher-level work or to those unfamiliar with the simulation's details. Also, the detailed models may be difficult and ponderous to use. The third group strongly prefers simple cause-effect models and the explanations they make possible. This group, however, is often not particularly interested in details and its simplified models may not be valid except in close-to-ideal circumstances, which makes them much less credible to organizations using the object model.

This disciplinary divide is itself an abstraction of reality and some individuals have their feet in more than one camp. Nonetheless, it seems to us that the divide is real and counterproductive. Motivated metamodeling is a way to build bridges among the camps. Table 4.1 suggests how the different classes of models might fare when compared along the various dimensions discussed.

Next Steps

The problem that we examined in this research was in some respects narrow. Significantly more thinking will be necessary to extend the ideas to other classes of problems—for example, when the object model is itself imperfect in some respects and other sources of knowledge exist about the truth; when only a small number of data points are available that represent behavior of the object model; when the object model is stochastic; when it is important to fine-tune the motivated metamodel for a particular analysis; or when analysts need computerized aids (including some of those used in data mining) to help them discover physical insights that could then be used to motivate metamodeling. Such topics would be appropriate subjects of further research.

Table 4.1

Comparison of Methods by Attributes of Resulting Models

Attribute	Statistical	Simulation	Reductionist	Motivated Metamodel
Story	●	● ●	● ● ● ●	● ● ● ●
Parsimony	● ●	●	● ● ● ●	● ● ●
Accuracy	● ● ●	● ● ● ●	● ●	● ● ●
Simplicity of explanation	●	● ●	● ● ● ●	● ● ●
Simplicity of use	● ● ●	● ●	● ● ● ●	● ● ●
Treatment of subtleties	●	● ● ● ●	●	● ● ●
Organizational acceptance	●	● ● ● ●	●	● ● ● ● [a]

[a] A reviewer suggested caution on this assessment. Organizational acceptance might not come easily.

Appendix

A. Implementing MRM with Switches

Figure A.1 illustrates implementing multilevel resolution with a "switch" using an example from freshman physics. The data-flow diagram in Figure A.1 is for a simulation that is computing a falling body's speed versus time, V(t). The simulation takes the previous time step's value and adjusts it by a term a(t) dependent on the forces of gravity and drag, g and D(t). Drag, however, may be represented simply as a constant D_{avg} or calculated (via a function F) from more detailed considerations that account for the object's shape and altitude, the weather, and so on. To implement this, we introduce a variable called switch, with values of "high" and "low." The calculation of V(t + Δt) would then be governed by something like the following:

$$\text{If switch} = \text{high}$$
$$\text{Then } D(t) = F\left(\text{object shape, etc.}\right)$$
$$\text{Else } D(t) = D_{avg}$$
$$V(t + \Delta t) = V(t) + \Delta t - D(t)\,\Delta t$$

This is trivial for a single example; a more complex multiresolution model could need a great many switches.

Although some ad hoc instances of such flexibility are common in existing models, it is unusual to have an entire model designed with multiresolution capabilities in mind. One reason is that the switches complicate the model's control structure and interface. Another reason is that the appropriate switches are sometimes context dependent and ought to be created as needed. In other words, which inputs are "natural" and "comfortable" depend on the applications and individuals involved. In most models, however, users are not provided with the flexibility to make their own choices. Instead, they are restricted to changing the values of inputs and interpreting the outputs (or requesting changes from a programming team, which may involve delays or frictions).[1]

---

[1]Fortunately, modern technology is making it easier to change aspects of a computer model other than just data. Spreadsheet technology is now ubiquitous, although not well suited to MRMPM, and some modeling systems, such as Analytica™ and iThink, have great inherent flexibility in high-level languages (McEver and Davis, 2001). For an ambitious recent example of MRMPM in Analytica, see Davis, McEver, and Wilson (2002).

38

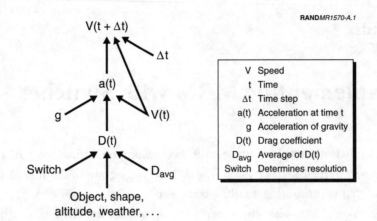

Figure A.1—Example of a Simple Switch

B. Details of the Experiment

This appendix is adapted from Bigelow and Davis (2002).

THE EXPERIMENT

Our experiment was to begin with a relatively large and complex model, EXHALT-CF, and to develop a series of metamodels to represent it. For the first metamodel, we relied almost entirely on simple statistical methods, uninformed by phenomenology (i.e., our knowledge of the workings of EXHALT-CF) and not going beyond standard day-to-day tools. With each successive metamodel, we took advantage of progressively more phenomenology.

EXHALT-CF, The Large Model

EXHALT-CF treats the halt phase of a military operation. In its simplest version, the halt phase is a mere race. An attacking force (Red) is advancing on an objective while the defenders (Blue) interdict Red's armored vehicles with long-range fires. Red will halt when he reaches his objective (a Red win) or when Blue has killed a specified number of vehicles (a Blue win), whichever comes first. EXHALT-CF, however, adds many embellishments relevant to current strategic concerns about real-world military operations, especially in the Persian Gulf (Davis, McEver, and Wilson, 2002).

First, the model must represent Blue deployments. Some number of shooters may be stationed in the theater in peacetime. Depending on strategic warning, diplomatic relations, Red's deceptiveness, and Red's ability to threaten bases in theater (e.g., with weapons of mass destruction), Blue may or may not be able to augment this number before Red begins his advance. Once Red's advance begins, Blue will deploy more shooters into the theater, up to a theater capacity, which reflects logistical shortcomings.

The effectiveness of Blue shooters is measured by kills per shooter-day. Early in the campaign, Blue may be unable or unwilling to attack the Red column because of Red air defenses. After a period of air-defense suppression, Blue's attacks will start. Even then, however, sortie rates may be reduced because of a continued

40

threat of attack with mass-destruction weapons, which would force Blue personnel to work in protective gear or to operate from more distant, more poorly prepared bases.

The weapons and strategy Blue selects will also influence Blue shooter effectiveness. Blue may select an area weapon capable of killing several Red armored vehicles per shot. To counter this, Red may space his vehicles more widely. Or Blue may select a point weapon, which kills no more than one vehicle per shot and is unaffected by Red's vehicle spacing. Also, Blue will likely have limited supplies of his best weapons and revert to lesser weapons when his best are exhausted. Blue may attack the entire Red column in depth (the In-Depth strategy) or focus his attack on the leading edge (the Leading-Edge strategy). If Blue does the latter, his attack may slow Red, but each sortie may be less effective because of deconfliction problems.

These and other complications of the halt problem are represented in EXHALT-CF. They are implemented in Analytica, a graphical modeling environment for the personal computer. EXHALT-CF has 63 inputs, 8 switches to turn features on or off (the model has a multiresolution, multiperspective design), three indexes, and 451 variables that are calculated directly or indirectly from inputs. For our purposes, we used a subset of the cases that the model can deal with, which reduced to 25 the number of input variables affecting the problem. This seemed adequately complex to illustrate our points.

The Experimental Data

We selected statistical distributions, mostly uniform distributions, from which to generate the 25 variable-value inputs.[1] We then ran EXHALT-CF to generate a Monte Carlo sample of 1000 cases from the overall input space and collected the variables shown in Table B.1. (We collected a few of the inputs that we held constant, as they will appear later in some of the equations.) We did not weight one or another region of the input space because we were seeking a broadly good fit of behavior over the entire domain of interest.

In some applications it is standard procedure to train the model on one dataset and validate it on a second dataset. This practice helps one avoid overfitting the model when the data are noisy. In our example, the object model, EXHALT-CF,

---

[1]We are aware, of course, that other sampling techniques such as the Latin Hypercube have advantages. Ideally, the experimental design should have nearly optimal properties for prediction under the model used, but with robustness to model misspecification. In our particular work, however, we were able to sample so extensively (1000 points) that it was not a major consideration.

Table B.1

Data Captured from EXHALT-CF

Variable	Description
VARIABLE INPUTS	
A^{00}	Initial (prewarning) shooters
Cday	C-day (no deception mean)
R0_frac	Fractional deployment rate at C-day without full access
Rstrat_frac	Fractional deployment rate on strategic warning
DT_strat	Extra (pre C-day) strategic warning
DT_access	Delay in access (post C-day)
R	Deployment rate
Amax	Theater capacity
Twait	Time until Blue starts attacking
Tsead	Time to suppress Red air defenses
Fpost	Post-wait antiarmor fraction
Delta_dep_area	Kills per shooter-day, area weapon, In-Depth strategy
Delta_le_area	Kills per shooter-day, area weapon, Leading-Edge strategy
Delta_dep_pt	Kills per shooter-day, point weapon, In-Depth strategy
Delta_le_pt	Kills per shooter-day, point weapon, Leading-Edge strategy
Delta_b	Kills per shooter-day, second-best ("B") weapon
AFV spacing	Armored fighting vehicle (AFV) spacing
N_awpns	Number of good ("A") weapons
Tdelay	Time before Red starts moving
Ndiv	Number of Red divisions
VPD	AFVs per Red division
H	Fraction AFVs to kill globally to stop Red
V	Red velocity, km/day
Hlocal	Fraction AFVs to kill locally to produce rout
Losses	Blue losses to air defense (Red air-defense kills)
CONSTANTS	
N_a	Number of "A" weapons per sortie
S_awpns	Sorties per shooter-day for "A" weapons
Obj	Red's objective
Axes	Number of axes for Red advance
Cols_per_axis	Columns per axis
OUTPUT VARIABLES	
Dhalt_dep_area	Distance Red achieves against area weapon, In-Depth strategy
Dhalt_le_area	Distance Red achieves against area weapon, Leading-Edge strategy
Dhalt_dep_pt	Distance Red achieves against point weapon, In-Depth strategy
Dhalt_le_pt	Distance Red achieves against point weapon, Leading-Edge strategy
Best_dhalt	Distance Red achieves against best strategy and weapon

was run in deterministic mode. There was no noise in the dataset to be mistaken for signal. Therefore, we use the entire dataset for fitting (i.e., training) the metamodels and reserve none of it for validation. In the absence of noise, the goodness-of-fit is adequate validation.

Metamodeling

Metamodel 1

In our first experiment, we acted as though we had handed the dataset to an average statistician (or statistically oriented operations researcher or computer scientist), and commissioned him to develop an estimator for *Best_dhalt*—the halt distance that Blue could achieve using his best strategy and weapon type for the circumstances of the case. A good statistician would prefer to have been involved in study design from the outset, but that is not typical. Nonetheless, even having been brought in late, the statistician would insist on discussing the problem with us. He would want to know which data elements to use as independent variables, which are outcome variables, and so on. Even if he preferred to operate as though the original model were a black box, he would probably want at least some interpretation of the variables' meanings, as indicated in the second column of the table. The statistician, then, would know that *Obj* is Red's objective, so that if Red achieves this distance, he will halt of his own volition. With this information, the statistician deduces that $0 \leq Best\_dhalt \leq Obj$ and then checks that all observations in the dataset satisfy these conditions. He understands that *Best_dhalt* will be insensitive to the variables listed in Table B.1 unless $0 < Best\_dhalt < Obj$. He will develop his estimator based on these cases and later check whether the estimator can be used to identify cases where *Best_dhalt=0* and *Best_dhalt=Obj*.

For his initial analysis, our statistician specifies a linear model with 25 independent variables (see Table B.2). Because he knows that we want a parsimonious metamodel, he runs a *forward stepwise selection* procedure in which the independent variables are added to the model one by one in the order of decreasing explanatory power. That is, the first variable considered yields the largest reduction of the root mean square of the residual error (RMS error). Figure B.1 shows that after the first six or seven variables, further additions do not improve the fit very much. Actually, in this case, the fit is not very good no matter how many variables are included. Although R^2 is about 0.6, which often passes for fairly good, the standard deviation of *Best_dhalt* in our dataset is about 160 km (taking only cases with $0 < Best\_dhalt < Obj$), and the RMS error never drops much below 100 km. Because halt distances for the domain of inputs

Table B.2

Variables in Metamodel 1

Variable	Coefficient	Std Dev	Product
Best_dhalt		160.45768	
Intercept	119.27582		
A^{00}	−0.43312	57.35259	24.84
Cday	6.56415	2.73085	17.93
R0_frac		0.08647	
Rstrat_frac		0.04335	
DT_strat		2.79181	
DT_access		2.87724	
R	−1.76637	8.59866	15.19
Amax	−0.05069	254.51938	12.90
Twait		0.59940	
Tsead	30.52580	1.62758	49.68
Fpost	−69.02157	0.14310	9.88
Delta_dep_area	−21.34313	3.12429	66.68
Delta_le_area		3.26664	
Delta_dep_pt		0.85954	
Delta_le_pt	−40.05037	0.88239	35.34
Delta_b	−53.93274	0.43084	23.24
AFV spacing		0.02840	
N_awpns	−0.00072	12608.00000	9.09
Tdelay	−41.67947	1.15126	47.98
Ndiv	3.49424	1.46579	5.12
VPD		79.32773	
H		0.12234	
V	3.49424	22.86022	79.88
Hlocal	302.86587	0.12234	37.05
Losses		7.77206	

range from 0 to 600 km and "good" halt distances are less than 100 km, the errors are significant to substantial.[2]

In any case, our simulated statistician stopped with 14 variables, all coefficients of which are significant at the 0.05 level. Table B.2, then, shows the coefficients, the standard deviations of the variables, and the product of the standard deviations and the absolute value of the coefficient. This product is a measure of the average effect the corresponding variable has on the best halt distance.[3]

[2]Shortcomings of pure forward stepwise selection are discussed in Rawlings (1988). They do not seem to apply here, because the input variables in the dataset were generated to be independent.

[3]The product, divided by the standard deviation, is the "standardized regression coefficient." That represents the impact of varying the particular variable away from its mean by a fixed fraction of its variance while holding all other variables at their mean values.

44

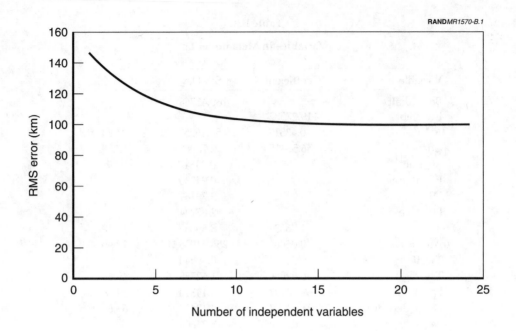

Figure B.1—RMS Error Versus Number of Variables for Model 1

Table B.2 indicates only how well Model 1 fits cases with $0<Best\_dhalt<Obj$, but these make up only 548 of the 1000 cases in our dataset. To extend Model 1 to the remaining cases, let L_1 be the linear function with coefficients as in Table B.2, and define:

$$M_1 = Max\{0, Min\{L_1, Obj\}\} \tag{B.1}$$

This, then, is Metamodel 1 (M_1).

How good is Metamodel 1? Earlier we identified four features that make a metamodel good. Table B.3 addresses goodness of fit when we apply the model M_1 to all 1000 cases in the dataset. The performance is not impressive, although in our experience with metamodeling, pure statistical approaches such as this not uncommonly do quite well by the average goodness-of-fit (or R^2) criterion. Still, in this case, there are 421 cases in which Red reaches his objective, and the model estimates that Red is halted short of his objective in 92 percent of them. The model's performance in these cases accounts for the large average underestimate of the halt distance (62.25 km).[4]

---

[4]A two-stage approach might have yielded a better metamodel. The first stage would involve developing a model to predict whether a particular input vector produces *Best_dhalt=0*, *Best_dhalt=Obj*, or an intermediate value. The second stage would develop a metamodel for the

Table B.3

Performance of Metamodel 1 on the Complete Dataset

Results	Value
$\text{Prob}(M_1 > 0 \mid \text{Best\_dhalt} = 0)$	0.52
$\text{Prob}(M_1 < \text{Obj} \mid \text{Best\_dhalt} = \text{Obj})$	0.92
$\text{Mean}(\text{Best\_dhalt} - M_1)$	62.25
$\text{RMS}(\text{Best\_dhalt} - M_1)$	140.02

Parsimony was our second criterion for a good metamodel. This model has 14 variables. We would like fewer, but this is perhaps a marginally acceptable number. We have found it possible to undertake comprehensive exploratory analysis with up to about a dozen variables (Davis, McEver, and Wilson, 2002).

Identification of important variables and critical components was our third criterion. Here Metamodel 1 performs very poorly. For example, from a logical point of view, *Losses* is just as important as A^{00} because adding one Blue shooter to A_{00} has the same effect as reducing *Losses* by one. But A_{00} is in the model (its coefficient is significant) while *Losses* is not (because the variation of A_{00} in our dataset is large and that of *Losses* is small). Similarly, *Ndiv* (number of divisions), *VPD* (vehicles per division), and *H* (break point) are all equally important, since they all enter EXHALT-CF only through their product xi (number of vehicles to kill for a halt). But of the three variables, only *Ndiv* appears in the model (because the percentage variation of *Ndiv* is large, while those of *VPD* and *H* are small). In short, regression has generated flatly wrong conclusions about the relative importance of variables. This could be a serious shortcoming if the model were used to inform resource-allocation decisions. Also, as discussed earlier, such metamodels are linear in the variables identified as significant by the statistical analysis. Thus, when used, Metamodel 1 fails to identify and highlight critical components, almost precisely as discussed earlier. For example, the model would predict that by merely improving munitions sufficiently, Blue could guarantee a short halt distance—independent of the other variables. Again, that is flatly wrong.

Our final criterion for a good metamodel was that it have a good storyl ine. This metamodel has no story line at all. It is what it is because "the computer says so."

---

intermediate cases. This approach would have made use of the information that we had about which combinations of inputs produced halt distances of zero or *Obj*. The approach we used ignored this information.

Metamodel 2

A statistician will often try to improve his model by introducing transformations of the independent variables, such as exponentials, powers, and products of variables. So many possible transformations of variables are possible that the statistician may need some guidance selecting which ones to try. Brute force (e.g., considering all of the quadratic combinations of elementary variables as new, composite variables) can result in a good fit, but usually with even more statistically significant variables and no "story."

Phenomenology (i.e., the "innards" of the baseline model) can suggest what transforms to try, including some that statisticians do not generally consider. These include transforms that use the Max and Min operators. Indeed, a number of transformations are built into EXHALT-CF. We designed it as a multiresolution model to permit the user to specify inputs at different levels of detail. Even if the metamodeler finds EXHALT-CF as a whole to be big and complex, usually early chapters of documentation, which deal with various idealizations, are sufficient to highlight natural composite variables. They may not *fully* substitute for the more elementary variables, because the "real" EXHALT-CF (as distinct from the simplified versions discussed in early documentation chapters) includes more complex interactions. Nonetheless, the suggested composite variables may go a long way.

For Metamodel 2, then, we looked at a number of such composites. First, A_0 (number of D-Day shooters) and xi (number of Red vehicles to kill for a halt) can be introduced as composite variables, which we have previously called aggregation fragments:

$$A_0 = f(A_{00}, Cday, R0\_frac, Rstrat\_frac, DT\_strat, DT\_acess) \tag{B.2}$$

$$\xi = Ndiv \times VPD \times H \tag{B.3}$$

The function defining A_0 is a bit tedious, so we do not write it out here, but it is readily understandable.

Next, the variables *Twait*, *Tsead*, and *Fpost* describe how Blue forces can be employed during the first few days of the campaign. They cannot be employed at all for the first *Twait* days. A fraction *Fpost* of Blue shooters can be employed between *Twait* and *Tsead*, and all Blue shooters can be employed after *Tsead*. If the halt time is larger than *Tsead*, we might reasonably replace all three variables by the single quantity *Tx*, where

$$Tx = Fpost \times Twait + (1 - Fpost) \times Tsead \tag{B.4}$$

We motivate this as follows. Consider running the model for two cases, one with the given values for *Twait*, *Tsead*, and *Fpost*, and the other with *Twait* and *Tsead* replaced by *Tx* (in this case, *Fpost* is moot). All other inputs to the two cases will be the same. We argue that the model will calculate virtually the same halt distance in the two cases, because the Blue shooters will contribute virtually the same number of shooter-days, and hence the same number of kills, between the start of the campaign and any time later than the first case's *Tsead*. (There will be a small difference if the number of Blue shooters in the theater changes between *Twait* and *Tsead*.) We will refer to *Tx* as the "equivalent wait time."

Next, Blue will not use both area weapons and point weapons in the same case. Blue always selects the better of the two. So we can define

$$Delta\_dep = Max\{Delta\_dep\_area, Delta\_dep\_pt\} \tag{B.5}$$

$$Delta\_le = Max\{Delta\_le\_area, Delta\_le\_pt\} \tag{B.6}$$

Indeed, we go beyond this and introduce the concept of "required shooter-days." This concept, which seldom is identified in a bottom-up approach to modeling, arises naturally in a more analytically oriented top-down approach of the sort encouraged in multiresolution modeling—or in a rough-cut attempt to understand what is going on analytically. It combines ξ (the number of Red vehicles to kill) and Blue effectiveness (*Delta_dep* and *Delta_le*) into a single quantity, and even incorporates the change in Blue effectiveness that occurs when Blue exhausts his good ("A") weapons. For the In-Depth strategy, we define

$$\xi\_dep\_A = Min\left\{\xi\left[\frac{N\_awpns \times Delta\_dep}{N\_a \times S\_awpns}\right]\right\} \tag{B.7}$$

Then, required shooter-days becomes:

$$RSD\_dep = \left[\frac{\xi\_dep\_A}{Delta\_dep}\right] + \left[\frac{\xi - (\xi\_dep\_A)}{Delta\_b}\right] \tag{B.8}$$

We define required shooter-days for the Leading-Edge strategy (*RSD_le*) similarly.

Now the statistician can define a linear model with far fewer variables, but the new variables include the effects of all the variables in M_1. The new metamodel, M_2 is developed using the 11 independent variables listed in Table B.4, but only ten of them prove significant at the 0.05 level in the final model (AFV spacing is insignificant).

Table B.4

Variables in Metamodel 2

Variable	Coefficient	Std Dev	Product
Best_dhalt		160.45768	
Intercept	14.48042		
A_0	−0.42042	57.35259	24.11
R	−2.27924	8.59866	19.60
Amax	−0.13580	254.51938	34.56
Tx	28.73035	1.31579	37.80
Tdelay	−39.05527	1.15126	44.96
Hlocal	104.18147	0.12234	12.75
RSD_dep	0.02447	1520.45869	37.21
RSD_le	0.05623	1499.80223	84.33
V	4.53483	22.86022	103.67
AFV spacing		0.02840	
Losses	2.19419	7.77206	17.05

As was the case for M_1, the linear function obtained by regression was estimated from the 548 cases in our dataset for which $0 < Best\_dhalt < Obj$. To extend Model 2 to the remaining cases, we denote by L_2 the linear function with the coefficients in Table B.4, and define

$$M_2 = Max\{0, Min\{L_2, Obj\}\} \qquad (B.9)$$

How good is Metamodel 2? When we apply the model M_2 to all 1000 cases in the dataset, we get the fit shown in Table B.5. The performance, while considerably better than Model 1, is still not impressive.

Because it has only ten independent variables, on grounds of parsimony it improves a bit on M_1.

We have built into this metamodel many nonlinear features that are necessary preconditions for identifying critical components. One is the construct "required shooter days," which combines the size of the threat and the effectiveness of Blue shooters in the appropriate nonlinear way. The job is not done, however. M_2's estimate of Red's halt distance depends on required shooter-days under both the

Table B.5

Performance of Metamodel 2 on the Complete Dataset

Results	Value
Prob($M_2 > 0$ | Best_dhalt=0)	0.68
Prob($M_2 < Obj$ | Best_dhalt=Obj)	0.32
Mean(Best_dhalt − M_2)	13.09
RMS(Best_dhalt − M_2)	83.56

In-Depth and Leading-Edge strategies, suggesting that reducing the required shooter-days needed under one strategy will reduce the distance Red advances when Blue chooses the other strategy. That, of course, is absurd, but the structure of Metamodel 2 does not include features to represent the nonsmooth surfaces generated by either-or choices. More generally, M_2 still implies that one can ensure a good halt distance by merely ratcheting one of the helpful variables sufficiently. In particular, ratcheting up RSD_dep or RSD_le sufficiently can substitute for even a very large Tx, whereas truth is otherwise.

Finally, this metamodel still does not have a story line, although the variables at least have more physical significance. In candor, we had expected that identifying these aggregation fragments would help more than it did.

Metamodel 3

So far the statistician has been combining the original, low-level variables from Table B.1 into intermediate variables that we think are reasonable on phenomenological grounds. We might characterize this as a bottom-up strategy. Now we turn to a top-down strategy. We view this as explicitly building in a story line. It depends on some understanding of aggregate phenomenology (it is certainly not for those who want to treat the original model as a black box), but does not require that the theory for describing that phenomenology be analytically tractable (e.g., solvable). It may be thought of as a reasonable guess about how factors would affect the problem, tidied up by a statistician.

In EXHALT_CF, Blue makes two decisions in each run: (a) whether to employ the area weapon or the point weapon, and (b) whether to use the In-Depth strategy or the Leading-Edge strategy. As explained earlier, Blue chooses the weapon with the larger kills per shooter-day (see Eqs. (B.5) and (B.6)). But it is not so easy to identify the better strategy. The In-Depth strategy attacks all parts of the Red column equally and assumes that Red advances at a constant velocity V. The Leading-Edge strategy attacks only the front of the Red column. Thus, in any time t, Red advances a distance Vt, less a "rollback" distance equal to the length of the column occupied by the Red vehicles killed during t.

Instead of estimating *Best_dhalt* directly, Metamodel 3 will estimate two distances, one for each strategy. We define them as

$$Best\_dhalt\_dep \ = \ Min\{Dhalt\_dep\_area, Dhalt\_dep\_pt\} \tag{B.10}$$

$$Best\_dhalt\_le \ = \ Min\{Dhalt\_le\_area, Dhalt\_le\_pt\} \tag{B.11}$$

Note in Table B.1 that the four distances *dhalt_dep_area*, *dhalt_dep_pt*, *dhalt_le_area*, and *dhalt_le_pt* are among the outcomes captured in our dataset of 1000 cases.

For the In-Depth strategy, we reason as follows. Let $A(t)$ be the number of Blue shooters in the theater at time t. Starting at D-Day, Blue will wait a time Tx (the equivalent wait time defined by Eq. (B.4)) without shooting, and then shoot continuously thereafter until the cumulative Blue shooter-days equals the required shooter-days (see Eq. (B.8)). Letting *Tshoot_dep* denote the length of time Blue continues shooting under the In-Depth strategy, we have

$$\int_{Tx}^{Tx+Tshoot\_dep} A(t)dt = RSD\_dep \qquad (B.12)$$

We approximate this integral by the product of a duration and an average number of shooters *Abar*. It is this kind of approximation that is possible without doing the detailed mathematics. It is more than dimensional analysis (i.e., ensuring consistent units from term to term), but not too much more. So our estimate of the duration will be $RSD\_dep/Abar$.

In EXHALT-CF, the number of Blue shooters is determined quite simply. There are A_0 shooters in the theater on D-Day, and R additional shooters can deploy into the theater each day thereafter. Between *Twait* and *Tsead* (which we have replaced by Tx), Blue loses *Losses* shooters. Finally, the number is never allowed to exceed the theater capacity, *Amax*. So for times after Tx we calculate the number of shooters as

$$A(t) = Min\{Amax, A_0 - Losses + R \times Tx + R \times (t - Tx)\} \quad for\ t \geq Tx \quad (B.13)$$

To keep it simple, we estimate *Abar* as the number of shooters in the theater when Red will have advanced a standard distance, say *Dstd*. Thus, for each case we estimate[5]

$$Abar \approx Min\left\{Amax, A_0 - Losses + R \times \left[Tdelay + \frac{Dstd}{V}\right]\right\} \qquad (B.14)$$

We provide ourselves with some degrees of freedom by introducing coefficients before each of the terms. Thus, we will fit the following linear function using only cases where *0<Best_dhalt_dep<Obj*:

$$L\_dep_3 = c_0 - c_1 \times V \times Tdelay + c_2 \times V \times Tx + c_3 \times V \times \frac{RSD\_dep}{Abar} \qquad (B.15)$$

---

[5]This is a typical kind of simplification, but not a very good one. Later, we discuss a much better estimator.

We develop a similar expression for the Leading-Edge strategy. It differs in that we must use *RSD_le* in place of *RSD_dep*, and that we must adjust the distance Red advances by the rollback distance. Fortunately, there is a simple and plausible estimate for the rollback distance in cases where *0<Best_dhalt_le<Obj*: the distance occupied by ξ, the number of Red vehicles to be killed. It should be exact except when Blue kills some vehicles before Red starts advancing (at *Tdelay*). The estimate of the rollback distance is

$$\Delta Droll = \frac{\dfrac{\xi}{H_{local}}}{\dfrac{(Cols\_per\_axis) \times Axes}{AFV\_spacing}} = \frac{AFV\_spacing \times \xi}{(Cols\_per\_axis) \times Axes \times H_{local}} \tag{B.16}$$

The numerator is the number of AFVs taken out of action if *vehicles are killed.* H_{local} is the local break point.

Then Model 3 for the Leading-Edge strategy becomes

$$L\_le_3 = c_0 - c_1 \times V \times Tdelay + c_2 \times V \times Tx + c_3 \times V \times \frac{RSD\_le}{Abar} - c_4 \times \Delta Droll \tag{B.17}$$

Once again we determine the coefficients c_0, c_1, c_2, c_3, and c_4 by regression, using only cases where *0<Best_dhalt_le<Obj*. We fit both Eqs. (B.15) and (B.17) for values of *Abar* calculated for a range of standard distances *Dstd*. For each *Dstd*, we define

$$M_3 = Max\{0, Min\{L\_dep_3, L\_le_3, Obj\}\} \tag{B.18}$$

We tried a range of values for *Dstd*, and found that *Dstd* = 400 gave good results. The coefficients of L_dep^3 and L_le^3 are shown in Tables B.6 and B.7, and the performance measures for M_3 in Table B.8. As shown there, this model improves substantially on Models 1 and 2. This is hardly surprising, since M_3 now contains a great deal of theoretical understanding of the problem. Still, it is a relief to see that the work paid off.[6]

---

[6]At this point we could test to determine which portions of the separate regressions can be made common (promising candidates in this example are the intercept, *VTx*, and *VTdelay*). Replacing these six calibration constants (three for each strategy) by three constants might degrade the fit only slightly, and would certainly improve the metamodel aesthetically.

Table B.6

Variables in L_dep₃

Variable	Coefficient	Std Dev	Product
Best_dhalt_dep		155.46531	
Intercept	59.81097		
VTx	0.77838	91.74328	71.41
VTdelay	−0.79763	84.23182	67.19
V(RSD_dep/Abar)	1.00632	142.54269	143.44

Table B.7

Variables in L_le₃

Variable	Coefficient	Std Dev	Product
Best_dhalt_le		161.64571	
Intercept	50.66498		
VTx	0.70943	94.64442	67.14
VTdelay	−0.70074	83.27428	58.35
V(RSD_le/Abar)	0.97397	154.11177	150.10
ΔDroll	−0.60021	24.45065	14.68

How good is Metamodel 3? Table B.8 shows that M_3 fits the data much better than either of the two previous metamodels. It is also much more parsimonious than the previous metamodels. If we consider rows in Tables B.6 and B.7 to represent independent variables, we count five, excluding the intercepts and duplicates. M_3 also highlights critical components, although this may not be immediately evident and the form of M_3 is still linear (in the composite variables). Because the coefficients of *VTx* and *V(RSD_dep/Abar)* are both positive, decreasing one cannot substitute for a sufficiently large value of the other. That is, keeping *both* terms small is necessary. Physically, this corresponds to saying that if Blue cannot start attacking the invading Red force early enough, he cannot achieve a good halt distance even if he has incredibly effective shooters. This is correct and much clearer than in M_2.

Finally, M_3 (unlike the previous metamodels) is based on a persuasive story. Some of the mathematics in Eqs. (B.10)–(B.18), although ultimately elementary, would be offputting to most clients, but the story can be explained adequately with something simpler (e.g., with all of the Max and Min operators removed, with more word equations rather than equations with symbols, and with a focus on only one case).

53

Table B.8

Performance of Metamodel 3 on the Complete Dataset

Results	Value
Prob(M_3>0 | Best_dhalt=0)	0.42
Prob(M_3<Obj | Best_dhalt=Obj)	0.01
Mean(Best_dhalt – M_3)	–11.51
RMS(Best_dhalt – M_3)	30.07

We judge M_3 to identify critical components very successfully. Simply by examination of Eqs. (B.15) and (B.17), we can see that if Red can move fast (V is large), it is essential for Blue to address three issues: (1) he must start shooting early (*Tx* must be small), (2) he must employ a large number of shooters (*Abar* must be large), and (3) he must limit required shooter-days. Referring back to Eqs. (B.3)–(B.8), limiting shooter-days can be accomplished by improving the effectiveness of weapons (including the provision of adequate numbers of "A" weapons).

Metamodel 4

Our final metamodel will be the same as M_3 except that we will use the exact solution *Tshoot_dep* to Eq. (B.12) (See Appendix C). Thus, we will fit the following linear function to cases where *0<Best_dhalt_dep<Obj*:

$$L\_dep_4 = c_0 - c_1 \times V \times Tdelay + c_2 \times V \times Tx + c_3 \times V \times Tshoot\_dep \quad \text{(B.19)}$$

We develop a similar expression for the Leading-Edge strategy:

$$L\_le_4 = c_0 - c_1 \times V \times Tdelay + c_2 \times V \times Tx + c_3 \times V \times Tshoot\_le - c_4 \times \Delta Droll \quad \text{(B.20)}$$

Once again we fit the function to cases where *0<Best_dhalt_le<Obj*. Then we define

$$M_4 = Max\{0, Min\{L\_dep_4, L\_le_4, Obj\}\} \quad \text{(B.21)}$$

The coefficients of L_dep_4 are shown in Table B.9, the coefficients of L_le_4 in Table B.10, and the performance measures for M_4 in Table B.11.

Table B.9

Variables in L_dep_4

Variable	Coefficient	Std Dev	Product
Best_dhalt_dep		155.46531	
Intercept	−2.43339		
VTx	1.02127	91.74328	91.94
VTdelay	−1.00265	84.23182	84.46
VTshoot_dep	1.01168	151.92006	153.69

Table B.10

Variables in L_le_4

Variable	Coefficient	Std Dev	Product
Best_dhalt_le		161.64571	
Intercept	2.87617		
VTx	0.95846	94.64442	90.71
VTdelay	−0.92003	83.27428	76.61
VTshoot_le	0.99000	161.27965	159.66
ΔDroll	−0.92681	24.45065	22.66

Table B.11

Performance of Metamodel 4 on the Complete Dataset

Results	Value
Prob($M_4 > 0$ | Best_dhalt=0)	0.03
Prob($M_4 <$ Obj | Best_dhalt=Obj)	0.01
Mean(Best_dhalt − M_4)	−0.01
RMS(Best_dhalt − M_4)	7.68

How good is Metamodel 4? As Table B.11 shows, M_4 fits the data almost perfectly.[7] It has the same number of independent variables as M_3, so it scores high on parsimony. And it is based on the same story as M_3.

M_4 incorporates the same critical components that we identified in M_3, but they are obscured by the mathematics. It requires a good deal of sophistication to explain, solely by inspection of the integral equation (B.12) or the linear equations using the variables of Tables B.9–B.10, how the critical components come into play. An analyst might feel comfortable presenting a client with a ratio (such as $RSD\_dep / Abar$ in M_3), and explaining that to reduce the ratio one can either

---

[7]This was not "supposed" to happen and should be regarded as quite unusual, certainly not something to be expected when dealing with object models a good deal more complex than EXHALT-CF.

reduce the numerator or increase the denominator. But how many analysts would present the client with an integral, much less try to explain it!

One reason for including M_4 in this report is to note that even a dedicated phenomenologist should be impressed by its performance compared to the very complicated original model. Many of the carefully derived features of EXHALT-CF do not ultimately pay their way.

Summary and Lessons Learned

Our experiments confirmed our hypothesis that much could be gained by combining virtues of statistical and phenomenological (theory-informed) approaches. They confirmed and give more precise arguments to our skepticism about approaching metamodeling as an exercise in pure data analysis, with the baseline model merely being a black-box generator of data. Although the experiment dealt with only a single baseline model, the insights appear to us to be relatively general—at least to suggest general cautions and approaches to consider.

In our experiments, the application of simple statistical methods uninformed by phenomenology did not produce a good metamodel. In part the application failed because the data we were fitting describe a highly nonlinear surface. A linear model might fit locally, but never globally. Moreover, there is no guarantee that the regression coefficients will be a good guide to the relative global importance of the independent variables.[8] It was necessary to introduce nonlinear combinations of the low-level inputs to obtain a good-fitting metamodel. Phenomenology can motivate the construction and selection of the appropriate nonlinear combinations. It would be very difficult to discover them from the data alone.

A linear model also fails because the data do not describe a smooth surface. Like many models, EXHALT-CF makes liberal use of Max and Min operators to make either-or choices. For example, Blue uses area weapons in some cases and point weapons in the rest. Blue employs the In-Depth strategy in some cases and the Leading-Edge strategy in others. Similarly, the number of "A" weapons is unimportant except in cases where they are exhausted. So the data describe a surface with "kinks" in it. When one fits a kinky surface with regression, the coefficients obtained from regression do not need to make sense. The regression

---

[8]Coefficients from a linear regression based on data from a local neighborhood will identify the relative importance of variables in that neighborhood, so long as the output is smooth enough (e.g., continuous).

results tell us the average importance of inputs relating to the different weapons and strategies, and to the number of "A" weapons available, but not *when* each one is important. We found it necessary to use phenomenology to separate the smooth segments of the surface, after which we could fit each smooth segment with just plain statistics.

As we introduced more and more phenomenology, we obtained successively better-fitting models. We would argue that in addition Model 1 had very little cognitive value (i.e., it did not tell a story) and Model 2 had only a little more. Model 3, however, does tell a coherent story, one that can be related persuasively to the client. With Model 4 it can be argued that we began to lose cognitive value. The phenomenology became more complex and less transparent. All the equations began to obscure the story. This is a subjective assessment, of course. Those with greater mathematical skill and more familiarity with EXHALT-CF might still see the story in a complex metamodel. But the client would not (nor, even, would other analysts, unless they invested a good deal of time).

Particularly relevant also is that by inserting phenomenologically motivated structure one can avoid certain blunders of system depiction. In particular, one can preserve and even highlight the role of critical components—components that enter the problem more nearly as products than as sums, or components that must individually have threshold values to avoid system failure. This is particularly significant if metamodels are to be used in policy analysis or design.

C. Exact Solution of Equations (14) and (16)

For reference we repeat Eqs. (B.12) and (B.13) here, except that we eliminate references to the strategy (In-Depth versus Leading Edge).

$$\int_{Tx}^{Tx+Tshoot} A(t)dt = RSD \tag{C.1}$$

$$A(t) = Min\{Amax, A_0 - Losses + R \times Tx + R \times (t - Tx)\} \quad for\ t \geq Tx \tag{C.2}$$

We define *Tmax* to be the time at which Blue shooters equal *Amax*. That is,

$$Tmax = \frac{Amax + Losses - A_0}{R} \tag{C.3}$$

For convenience we also define

$$Ax = A_0 - Losses + R \times Tx \tag{C4}$$

Then there are three cases to consider, namely $Tmax \leq Tx$, $Tx < Tmax < Tx + Tshoot$, and $Tmax \geq Tx + Tshoot$.

If Tmax<Tx

In this case, $A(t) = Amax$ for all $t \geq Tx$. So

$$Tshoot = \frac{RSD}{Amax} \tag{C.5}$$

If Tx < Tmax < Tx + Tshoot

In this case, $A(t)$ will have reached the theater capacity *Amax* before Blue has finished shooting. The total shooter-days accumulated between *Tx* and *Tmax* is easily calculated as $0.5 \times (Ax + Amax) \times (Tmax - Tx)$. After *Tmax*, the number of Blue shooters will be a constant, *Amax*. So the time it takes Blue to accumulate *RSD* shooter-days will be

$$Tshoot = (Tmax - Tx) + \frac{RSD - 0.5 \times (Ax + Amax) \times (Tmax - Tx)}{Amax} \tag{C.6}$$

After you perform this calculation, check whether $Tmax < Tx + Tshoot$. If so, this is the correct equation to use. Otherwise, use the third case.

If Tmax ≥ Tx + Tshoot

In this case, Blue will finish shooting before $A(t)$ reaches the theater capacity $Amax$. Since $A(t)$ increases linearly during this period, its integral increases as a quadratic, and $Tshoot$ is the solution to that quadratic:

$$Tshoot = \frac{-Ax + \sqrt{Ax^2 + 2 \times R \times RSD}}{R} \tag{C.7}$$

D. Measuring Goodness of Fit

We are given data points $(y_j, x_{j1}, \ldots, x_{jm})$ for $j = 1, 2, \ldots, n$. The m x-variables are the independent variables in our metamodel, while y is the dependent variable. The metamodel is

$$y \approx M(x_1, x_2, \ldots, x_m; c_0, c_1, c_2, \ldots, c_k) \tag{D.1}$$

The parameters $c_0, c_1, c_2, \ldots$, ck are calibration parameters, to be adjusted so Eq. (D.1) fits the given data points as well as possible.

We can use Eq. (D.1) to calculate an error for every data point. Thus, for point j we have

$$\varepsilon_j = y_j - M(x_{j1}, x_{j2}, \ldots, x_{jm}; c_0, c_1, c_2, \ldots, c_k) \tag{D.2}$$

Standard statistical regression procedures choose values for the $c_0, c_1, c_2, \ldots, c_k$ that drive the sum of squares of the errors j to a minimum. Thus they are using the sum of squares of the errors as a measure of fit, which gets better as it gets smaller. The sum of the squares of the errors is difficult to interpret, however, so statistical software packages generally include two transformed versions of this quantity in the outputs to their regression routines. Ignoring pesky corrections for the numbers of degrees of freedom (which are small if the number of observations, n, is large), the two quantities are

$$RMSE = \sqrt{\frac{\sum_j \varepsilon_j^2}{n}} \tag{D.3}$$

$$R^2 = 1 - \frac{RMSE^2}{\sigma_y^2} \tag{D.4}$$

The quantity σ_y is the standard deviation of the dependent variable y. If the average of the errors ε_j is zero (which is guaranteed if the model M is linear in the calibration parameters $c_0, c_1, c_2, \ldots, c_k$, and if one of the parameters, say c_0, is an intercept), then RMSE (which stands for root mean square error) is the standard deviation of the error terms.

Even though RMSE and R^2 are mathematically equivalent, once σ_y is known, we prefer RMSE over R^2 as a measure of the goodness of fit, for two reasons. First, RMSE is directly comparable to y, because it is in the same units. In our example, both RMSE and y are distances. Thus, RMSE is a direct measure of how badly the metamodel estimates the distance Red penetrates. But R^2 scales this discrepancy by the dispersion σ_y within the data points of the Red penetration distances. This makes it hard to judge how well the metamodel does. Second, the fact that the dispersions are squared in R^2 can be misleading. An R^2 of 0.75 is often regarded as evidence of a very good fit indeed, yet in our example the corresponding RMSE is 80 km (in our example, σ_y = 160 km). Even an R^2 of 0.9, very high indeed, corresponds to an RMSE of over 50 km. The problem is that to calculate RMSE from R^2, one uses the square root function, which is very steep when R^2 is near 1 (infinitely steep in the limit).

Many statistical packages allow the user to weight the data points unequally. The regression procedure then finds the calibration parameter values that minimize the weighted sum of the squared errors. In the resulting metamodel, the errors at points with high weights will tend to be smaller than the errors in the unweighted model. In compensation, the errors at points with low weights will be larger. In other words, weights merely redistribute the errors, generally without making the overall fit better. There is no free lunch. On the other hand, if the analyst considers it significant that the model fit certain regions well, weighting is a way to accomplish it.[1]

For example, if the weight on point j is $(1/y^{j2})$, then the regression procedure will minimize the sum of squares of fractional errors. Making this choice implies that a 10 percent error in the estimate of Red's penetration distance is equally serious whether it is 10 percent of 20 km or 10 percent of 200 km.

We do not recommend minimizing one measure of the goodness of fit to determine the values of the calibration parameters, and using a different measure to assess the goodness of fit of the resulting metamodel. The metamodel generated by an unweighted regression will rate a 10-km error in a penetration distance of 20 km (a 50 percent error) as no more significant than a 10-km error in a penetration distance of 200 km (a 5 percent error). So errors for data points with small y-values are likely to be about as large as errors for points with large y-values. If the analyst finds this a shortcoming of the metamodel, he should have chosen different weights.

---

[1]Much the same effect is produced by oversampling from key regions, i.e., using the object model to generate more points from more important regions and fewer from less important regions.

Bibliography

An, Jian, and Art Owen (2001), "Quasi-Regression," *Journal of Complexity,* Vol. 17, No. 4, pp. 588–607.

Axtell, Robert (1992), "Theory of Model Aggregation for Dynamical Systems with Applications to Problems of Global Change," Dissertation, Carnegie-Mellon University, Pittsburgh, PA. Available from University Microfilms International, Ann Arbor, MI.

Balci, Osman (1994), "Validation, Verification, and Testing Techniques Throughout the Life Cycle of a Simulation Study," *Annals of Operations Research,* Vol. 53, pp. 121–173.

Bankes, Steven C. (1993), "Exploratory Modeling for Policy Analysis," *Operations Research,* Vol. 41, No. 3.

Bankes, Steven C. (2002), "Tools and Techniques for Developing Policies for Complex and Uncertain Systems," *Proceedings of the National Academy of Sciences,* Colloquium, Vol. 99, Suppl. 3, pp. 7263–7266.

Bellman, Kirstie (1990), "The Modeling Issues Inherent in Testing and Evaluating Knowledge-Based Systems," *Expert Systems with Applications Journal,* Vol. 1, pp. 199–215.

Bigelow, James H., and Paul K. Davis (2002), "Developing Improved Metamodels by Combining Phenomenological Reasoning with Statistical Methods" *Proceedings of the SPIE,* Vol. 4716, pp. 167–180.

Box, G. E., and N. R. Draper (1987), *Empirical Model-Building and Response Surfaces,* John Wiley, NY.

Cassandras, Christopher et al. (2000), *Stochastic Fidelity Preservation in Mixed Resolution Simulation Modeling,* Network Dynamics, Burlington, MA.

Cloud, David S., and Larry B. Rainey (1995), *Applied Modeling and Simulation: An Integrative Approach to Development and Operations,* McGraw Hill, NY.

Cressie, Noel A. C. (1993), *Statistics for Spatial Data,* revised edition, John Wiley, New York.

Davis, Paul K. (1993), *An Introduction to Variable Resolution Modeling and Cross-Model Connection,* RAND, R-4252. Also published in *Journal of Defense Logistics* and reprinted as a chapter in Jerome Bracken et al., *Combat Modeling.*

Davis, Paul K. (1994), *New Challenges for Defense Planning: Rethinking How Much Is Enough,* RAND, MR-400-RC.

Davis, Paul K. (2001a), "Exploratory Analysis Enabled by Multiresolution, Multiperspective Modeling,"" in Jeffrey A. Joines, Russell R. Barton, K. Kang, and Paul A. Fishwick (eds.), *Proceedings of the 2000 Winter Simulation Conference,* 2000. Available from RAND as RP-925.

Davis, Paul K. (2001b), *Effects-Based Operations: A Grand Challenge for Analysis,* RAND, MR-1477, USJFCOM/AF.

Davis, Paul K. (2001dc), "Selected Comments on Defining and Measuring Machine Intelligence," in NIST (2001).

Davis, Paul K. (2002a), "Synthetic Cognitive Modeling of Adversaries for Effects-Based Planning," *Proceedings of the SPIE,* Vol. 4716 (edited by Alex Sisti and Dawn Trevasani).

Davis, Paul K. (2002b), *Analytic Architecture for Capabilities-Based Planning, Mission-System Analysis, and Transformation,* RAND, MR-1513-OSD.

Davis, Paul K., and James H. Bigelow (1998), *Experiments in Multiresolution Modeling,* RAND, MR-1004-OSD.

Davis, Paul, and James H. Bigelow (2001), "Metamodels to Aid Planning of [by] Intelligent Machines," *Proceedings of PerMIS 2001,* Mexico City, National Institutes of Standards and Technology, Gaithersberg, MD.

Davis, Paul K., James H. Bigelow, and Jimmie McEver (1999), *Analytical Methods for Studies and Experiments on "Transforming the Force,"* RAND, DB-278-OSD.

Davis, Paul K., James H. Bigelow, and Jimmie McEver (2001a), *Effects of Terrain, Maneuver Tactics, and C4ISR on the Effectiveness of Long Range Precision Fires: A Stochastic Multiresolution Model (PEM) Calibrated to High-Resolution Simulation,* RAND, MR-1138-OSD

Davis, Paul K., James H. Bigelow, and Jimmie McEver (2001b), *Exploratory Analysis and a Case History of Multiresolution, Multiperspective Modeling,* RAND, RP-925 (a reprint volume drawing on articles from the 2000 Winter Simulation Conference and Proceedings of the SPIE).

Davis, Paul K., David Gompert, and Richard Hillestad (1996), *Adaptiveness in National Defense: Suggestions for a New Framework,* RAND, Issue Paper IP-155.

Davis, Paul K., and Richard Hillestad (1993), "Families of Models that Cross Levels of Resolution: Issues for Design, Calibration and Management," *Proceedings of the 1993 Winter Simulation Conference,* Los Angeles, CA.

Davis, Paul K., Richard Hillestad, and Natalie Crawford (1997), "Capabilities for Major Regional Conflicts," in Zalmay M. Khalilzad and David A. Ochmanek (eds.), *Strategic Appraisal 1997: Strategy and Defense Planning for the 21st Century,* RAND, MR-826-AF.

Davis, Paul K., and Robert Howe (1988), *The Role of Uncertainty in Assessing the NATO-Pact Central Region Balance,* RAND, N-2839.

Davis, Paul K., Jimmie McEver, and Barry Wilson (2002), *Measuring Interdiction Capabilities in the Presence of Anti-Access Strategies: Exploratory Analysis to Inform Adaptive Strategy for the Persian Gulf*, RAND, MR-1471-AF.

Davis, Paul K., and James A. Winnefeld (1983), *The RAND Strategy Assessment Center: An Overview and Interim Conclusions About Utility and Development Options*, RAND, R-2945-DNA.

Easterling, Robert G. (1999), *A Framework for Model Validation*, Sandia National Laboratories.

Fishwick, Paul K. (1995), *Simulation Model Design and Execution: Building Digital Worlds*, Prentice Hall, Englewood Cliffs, NJ.

Haimes, Yacov Y. (1998), *Risk Modeling, Assessment, and Management*, Wiley, NY.

Handock, Mark S., and Michael L. Stein (1993), "A Bayesian Analysis of Kriging," *Technometrics*, Vol. 35, pp. 403–410.

Hillestad, Richard, and Paul K. Davis (1998), *Resource Allocation for the New Defense Strategy: The DynaRank Decision Support System*, RAND, MR-996-OSD.

Hodges, James S. (1987), "Uncertainty, Policy Analysis, and Statistics," *Statistical Science*, Vol. 2, No. 3, August, pp. 259–291.

Kelton, W. David (1999), "Statistical Analysis of Simulation Output," *Proceedings of the 1999 Winter Simulation Conference*, Squaw Peak, AZ.

Kennedy, M. C., and A. O'Hagan (2000), "Predicting the Output from a Complex Computer Code When Fast Approximations Are Available," *Biometrica*, Vol. 87, No. 1, pp. 1–13.

Kleijnen, Jack P. C. (1999), "Validation of Models Statistical Techniques and Data Analysis," *Proceedings of the 1999 Winter Simulation Conference* (Analysis Methodology Track), Squaw Peak, AZ.

Landauer, Christopher (1990), "Correctness Principles for Rule-Based Expert Systems," *Expert Systems with Applications Journal*, Vol. 1, pp. 291–316.

Law, Averill M., and W. David Kelton (1991), *Simulation Modeling and Analysis*, second edition; McGraw-Hill, NY.

Lempert, Robert J. (2002), "A New Decision Science for Complex Systems," *Proceedings of the National Academy of Sciences*, Colloquium, Vol. 99, Suppl. 3, pp. 7309–7313.

Lund J., R. Berg, and C. Replogle (1993), *An Assessment of Strategic Airlift Operational Efficiency*, RAND, R-4269/4-AF.

McEver, Jimmie, and Paul K. Davis (2001), "Defining the Next Generation of Enabling Technology for Exploratory Analysis and Multi-Resolution Modeling," *Proceedings of the SPIE*, April.

McEver, Jimmie, Paul K. Davis, and James H. Bigelow (2000), *EXHALT: An Interdiction Model for Exploring the Interdiction Problem in a Large Scenario Space*, RAND, MR-1137-OSD.

McGraw, Robert (1998), *Model Abstraction Demonstration*, Final Report BAA 98-02-1FKPA, RAM Laboratories, San Diego, CA.

Messina, Elena, and Alex Meystel (2000), *Proceedings of PerMIS 2000*, National Institute of Standards and Technology, Gaithersberg, MD.

Meystel, Alex (2001), "Measuring Performance of Systems with Autonomy: A White Paper for Explaining Goals of the Workshop" in NIST (2001).

Meystel, Alex (1995), *Semiotic Modeling and Situation Analysis, an Introduction*, AdRem Inc., Bala Cynwyd, PA.

Meystel, Alex, and James Albus (2002), *Intelligent Systems: Architecture, Design, and Control*, John Wiley, NY.

Morgan, Granger, and Max Henrion (1990), *Uncertainty: A Guide to Dealing with Uncertainty in Quantitative Risk and Policy Analysis*, Cambridge University Press, reprinted in 1998.

Morton, Sally, and John E. Rolph (2000), "Racial Bias in Death Sentencing: Assessing the Statistical Evidence," in their edited volume, *Public Policy and Statistics*, Springer, NY.

National Institute of Standards and Technology (NIST) (2001), *Performance Metrics for Intelligent Systems* (PerMIS 2000), Gaithersburg, MD. Available on web.

National Research Council (1997), *Modeling and Simulation*, Vol. 9 of *Technology for the United States Navy and Marine Corps: 2000–2035*, Naval Studies Board, National Academy Press, Washington, D.C.

Network Dynamics, Inc. (2000), *Stochastic Fidelity Preservation in Mixed Resolution Simulation Modeling*, Final Report, SBIR Contract F30602-99-C0097, study done for Air Force Research Laboratory, January 15. C. G. Cassandras, principal author.

O'Hagan, A., Kennedy, M. C., and J. E. Oakley (1999), "Uncertainty Analysis and Other Inference Tools for Complex Computer Codes," in J. M. Bernardo, J. O. Berger, A. P. David, and A. F. M. Smith (eds.), *Bayesian Statistics 6*, Oxford University Press, London.

Pace, Dale K. (1998), "Verification, Validation, and Accreditation," Chapter 11 in Cloud and Rainey (1995).

Rawlings, John O. (1988), *Applied Regression Analysis: A Research Tool*, Wadsworth and Brooks/Cole, Pacific Grove, CA.

Reshansky, Alex, and Robert M. McGraw (2002), "MRMaide: A Mixed Resolution Modeling Aide," *Proceedings of the SPIE*, Vol. 4716 (ed. by Dawn Trevasani and Alex Sisti), RAM Labs., Inc.

Sacks, Jerome, WIlliam J. Welch, Toby J. Mitchell, and Henry P. Wynn (1989), "Design and Analysis of Computer Experiments," *Statistical Science*, Vol. 4, No. 4, pp. 409–435.

Saltelli, Andre (2002), "Making Best Use of Model Evaluations to Compute Sensitivity Indices," *Computer Physics Communications*, Vol. 145, No. 2, pp. 280–297.

Saltelli, Andrea, Karen Chan, and E. Marian Scott (eds.) (2000), *Sensitivity Analysis*, John Wiley and Sons, Chichester.

Sargent, Robert G. (1996), "Verifying and Validating Simulation Models," *Proceedings of the 1996 Winter Simulation Conference*, Squaw Peak, AZ.

Sisti, Alex, and A. Farr (1999) "Abstraction Techniques: An Intuitive Overview" *Proceedings of the SPIE*.

Sterman, John D. (2000), *Business Dynamics: Systems Thinking and Modeling for a Complex World*, Irwin McGraw Hill, Boston, MA.

Treshansky, Alan, and Robert McGraw, "An Overview of Clustering Algorithms," *Proceedings of the SPIE*, Aerosense, 2001.

Wright, Samuel A. (2001), *Covalidation of Dissimilarly Structured Models*, United States Air Force Institute of Technology, Wright Patterson Air Force Base, OH.

Ziegler, Bernard, Herbert Praenhofer, and Tag Gon Kim (2000), *Theory of Modeling and Simulation: Integrating Discrete Event and Continuous Complex Dynamic Systems*, second edition, John Wiley, San Diego, CA.